W9-BBO-903

Translations

of

Mathematical Monographs

Volume 38

Statistical Sequential Analysis

Optimal Stopping Rules

by

A. N. Širjaev

American Mathematical Society
Providence, Rhode Island
1973

СТАТИСТИЧЕСКИЙ ПОСЛЕДОВАТЕЛЬНЫЙ АНАЛИЗ
ОПТИМАЛЬНЫЕ ПРАВИЛА ОСТАНОВКИ

А. Н. ШИРЯЕВ

Издательство „Наука"
Главная Редакция
Физико-Математической Литературы
Москва 1969

Translated from the Russian by
Lisa and Judah Rosenblatt

AMS (MOS) subject classifications (1970).
Primary 62L12, 62L15; Secondary 60G40.

Library of Congress Cataloging in Publication Data

Shiriaev, Al'bert Nikolaevich.
Statistical sequential analysis.

(Translations of mathematical monographs, v. 38)
Translation of Statisticheskiĭ posledovatel'nyĭ analiz.
1. Sequential analysis. I. Title. II. Series.
QA279.7.S5213 519.5'4 73-4451
ISBN 0-8218-1588-1

Printed in the United States of America

TABLE OF CONTENTS

PREFACE

1. Sequential analysis as a method of statistical research became widely known after the appearance[1] (in 1947) of the book *Sequential analysis* by A. Wald. In this book he set forth the theory and gave applications for one particular procedure in sequential analysis, the so-called sequential probability ratio test. The importance of this procedure, as discovered by Wald and proved by him jointly with J. Wolfowitz **[73]** was that in discriminating between two simple hypotheses it minimizes the expected number of observations compared with any other test which has the same probabilities of wrong decisions.

In contrast to the classical method of distinguishing between two simple hypotheses (the Neyman-Pearson method), for which the number of observations is fixed in advance, in sequential analysis the sample size is random and its determination depends on the values of the observed data. The Wald sequential probability ratio test turned out to be the method that allows the time at which observation is terminated to be determined in an optimal fashion.

A detailed investigation of the questions of existence and ways of finding optimal stopping times in Bayesian decision procedures is given in well-known works by A. Wald and J. Wolfowitz **[74]**, and K.J. Arrow, D. Blackwell and M.A. Girshick **[3]**. The following general problem on optimal stopping of random processes with discrete time was stated shortly after the appearance of these works and under their influence by J.L. Snell **[68]**.

Suppose that on some finite probability space[2] $(\Omega, \mathscr{F}, \mathbf{P})$ we are given a nondecreasing sequence of σ-algebras $\mathscr{F}_0 \subseteq \mathscr{F}_1 \subseteq \cdots \subseteq \mathscr{F}_n \subseteq \mathscr{F}$ and a sequence of random variables $Z_n = Z_n(\omega)$, $n = 0, 1, \cdots$, which are $\mathscr{F}_n$-measurable for each n.

Let $\mathfrak{M} = \{\tau\}$ denote the collection of random variables $\tau = \tau(\omega)$ taking on the values $0, 1, \cdots$ and satisfying the condition $\{\omega: \tau(\omega) = n\} \in \mathscr{F}_n$ for each $n = 0, 1, \cdots$.

Such random variables are called *stopping times,* and we say that they determine a stopping rule.

We interpret Z_n as the "reward" obtained by stopping observation at time n, and we interpret $\mathbf{E}Z_\tau$ as the expected reward, corresponding to the stopping rule

[1] The Russian translation **[71]** appeared in 1960.

[2] The basic probability-theoretic concepts are presented in Chpater I.

$\tau = \tau(\omega)$. The basic problems in the theory of optimal stopping rules consist of finding the value $S = \sup_{\tau \in \mathfrak{M}} \mathbf{E}Z_\tau$ and ε-optimal times τ_ε, i.e. those stopping times for which $\mathbf{E}Z_{\tau_\varepsilon} \geqq S - \varepsilon$, $\varepsilon \geqq 0$. 0-optimal stopping times are simply called *optimal*.

On the basis of martingale theory, J.L. Snell showed (under certain assumptions) that for a sequence $\{Z_n\}$, $n = 0,1,\cdots$, there exists a minimal regular super-martingale $(Y_n,\mathscr{F}_n)$, $n = 0,1,\cdots$, majorizing Z_n, knowledge of which allows us to solve the problems posed above.

It turned out that the value $S = \mathbf{E}Y_0$ and the stopping time (under broad assumptions) $\tau_\varepsilon = \inf\{n \geqq 0: Y_n \leqq Z_n + \varepsilon\}$ is ε-optimal, $\varepsilon > 0$.

Further developments of Snell's results were given by Y.S. Chow and H. Robbins **[15]**—**[17]**, G.W. Haggstrom **[39]**, D.O. Siegmund **[56]** and others.

In the framework of the scheme formulated above, the most intuitive case is the one where the variables Z_n can be represented in the form $Z_n = g_n(\xi_0,\cdots,\xi_n)$, where $\xi_0,\xi_1,\cdots$ is a sequence of (observable) variables, and $\mathfrak{F}_n$ is the σ-algebra of ω-sets generated by $\xi_0,\cdots,\xi_n$. Of basic interest from the point of view of theory as well as applications is the case where the sequence $\xi_0,\xi_1,\cdots$ is Markovian. Indeed this case, first investigated by E.B. Dynkin in **[26]**, is the subject of study in the present book.

2. The book consists of four chapters. The first chapter is of a supportive nature. Here we recall the basic probability-theoretic concepts and present derivations in the theory of Martingales and Markov processes for later use. In the second section we introduce the concepts of Markov times and stopping times, and examine their properties.

The second and third chapters deal with the problems of existence, and ways of constructing ε-optimal and optimal Markov times for the discrete and continuous time cases respectively.

We examine in some detail the contents of the second chapter.

Let $X = (x_n,\mathscr{F}_n,\mathbf{P}_x)$, $n = 0,1,\cdots$, be a discrete time Markov process with phase space $(E,\mathscr{B})$. We use $\overline{\mathfrak{M}} = \{\tau\}$ to denote the class of Markov times (M.t.)[3] $\tau = \tau(\omega)$ (relative to the system of σ-algebras $\{\mathscr{F}_n\}$) taking on the values $0,1,\cdots,+\infty$. Further, let $\mathfrak{M} \subseteq \overline{\mathfrak{M}}$ be the class of M.t.'s finite with probability one $(\mathbf{P}_x(\tau < \infty) = 1$ for all $x \in E)$. Such M.t.'s are called *stopping times* (s.t.).

We say that the $\mathscr{B}$-measurable function $g(x)$ belongs to the class L if $\mathbf{E}_x g^-(x_n) < \infty$ for all $n = 0,1,\cdots$ and all $x \in E$, where $g^-(x) = -\min(g(x),0)$. The collection of functions $g(x) \in L$ satisfying the condition

$$A^-: \mathbf{E}_x\left[\sup_n g^-(x_n)\right] < \infty, \qquad x \in E,$$

is denoted by $L(A^-)$. Analogously, if the condition

[3] See Definition 1 on p. 8.

$$A^+ : \mathbf{E}_x \left[\sup_n g^+(x_n)\right] < \infty, \qquad x \in E,$$

is satisfied, $g^+(x) = \max\big(g(x),0\big)$, then we write $g(x) \in L(A^+)$. We also let $L(A^-,A^+) = L(A^-) \cap L(A^+)$.

Suppose

$$\mathbf{E}_x g(x_\tau) = \int_{\{\tau<\infty\}} g(x_\tau) d\mathbf{P}_x$$

and

$$\tilde{\mathbf{E}}_x g(x_\tau) = \int_{\{\tau<\infty\}} g(x_\tau) d\mathbf{P}_x + \int_{\{\tau=\infty\}} \limsup_n g(x_n) d\mathbf{P}_x.$$

We call each of the functions

$$s(x) = \sup_{\tau \in \mathfrak{M}} \mathbf{E}_x g(x_\tau)$$

and

$$\tilde{s}(x) = \sup_{\tau \in \bar{\mathfrak{M}}} \tilde{\mathbf{E}}_x g(x_\tau)$$

a "value". The time $\tau_\varepsilon \in \mathfrak{M}$ is said to be (ε,s)-optimal if $\mathbf{E}_x g(x_{\tau_\varepsilon}) \geqq s(x) - \varepsilon$ for all $x \in E$. The time $\tau_\varepsilon \in \bar{\mathfrak{M}}$ is said to be $(\varepsilon,\tilde{s})$-optimal if $\tilde{\mathbf{E}}_x g(x_{\tau_\varepsilon}) \geqq \tilde{s}(x) - \varepsilon$ for all $x \in E$.

In Theorem 1 it is shown that if the function $g(x) \in L(A^-)$, then the value $s(x)$ is the smallest excessive majorant of the function $g(x)$, i.e. the smallest of the functions $f(x) \in L(A^-)$ satisfying the conditions

$$f(x) \geqq g(x),$$
$$f(x) \geqq Tf(x), \quad Tf(x) = \mathbf{E}_x f(x_1).$$

It is also shown that the value $\tilde{s}(x)$ coincides with $s(x)$. In other words, widening the class of stopping times $\mathfrak{M}$ (to the class $\bar{\mathfrak{M}}$) does not lead to an increase in value.

The structure of (ε,s)- and $(\varepsilon,\tilde{s})$-optimal times (under the assumption $g \in L(A^-, A^+)$) is investigated in § 2. Here we also consider various examples in which we find the value and (ε,s)- or $(\varepsilon,\tilde{s})$-optimal times.

Discarding the assumption $g(x) \in L(A^-)$ leads, in general, to the situation that the value is not the smallest excessive majorant. In Theorem 3, however, it is shown that even in the case $g(x) \in L$ the functions $s(x)$ and $\tilde{s}(x)$ coincide, and the value is the smallest *regular* excessive majorant of the function $g(x)$ (see the definition in § 3).

In Theorem 4 we give conditions for existence of (ε,s)- and $(\varepsilon,\tilde{s})$-optimal times, $\varepsilon \geqq 0$, under the assumption that the function $g(x) \in L(A^+)$.

The fourth section deals with investigation of problems on optimal (i.e. $(0,s)$-optimal) stopping under the assumption that the given stopping times τ belong to the class $\mathfrak{M}_N$ $\big(\mathbf{P}_x(\tau \leqq N) = 1, x \in E\big)$. It is shown that for $N < \infty$ optimal stopping times τ_N^* exist and it is determined when $\lim_{N\to\infty} \tau_N^*$ is $(0,s)$- or $(0,\bar{s})$-optimal in the classes $\mathfrak{M}$ and $\overline{\mathfrak{M}}$.

The smallest excessive majorant $v(x)$ of the function $g(x)$ satisfies the recursion equation

$$v(x) = \max\big(g(x),\ Tv(x)\big)$$

(Lemma 3). In § 5 we investigate questions of a unique solution of these recursion equations.

In the sixth section we study the problem of when the optimal stopping time $\tau^* \in \mathfrak{M}$ is bounded,[4] i.e. there exists $N < \infty$ such that $\tau^* \in \mathfrak{M}_N$.

The seventh section is concerned with an investigation of sufficient and randomized classes of stopping times. In § 8 we consider the problems of optimal stopping under the assumptions that $\mathbf{E}_x g(\tau, x_\tau)$ or $\mathbf{E}_x[\alpha^\tau g(x_\tau) - \sum_{s=0}^{\tau-1} \alpha^s c(x_s)]$ are maximized.

In the third chapter problems on optimal stopping are studied for the case of (standard) continuous time Markov processes. A large number of the results obtained in this chapter are at least superficially similar to the corresponding results for the case of discrete time.

It should, however, be mentioned that in this chapter rather complex techniques of the theory of continuous time Markov processes are involved in the investigation of optimal stopping problems. Therefore in this chapter (in contrast to the second chapter) there are many references to the monographs of E.B. Dynkin **[24]**, **[25]** (and also of P.A. Meyer **[49]**, **[50]**, and R.M. Blumenthal and R.K. Getoor **[12]**).

In the fourth chapter we show how the theory of optimal stopping rules is applied to solving two problems in mathematical statistics: testing two simple hypotheses and the problem of "disruption". We touched on the first problem briefly above. In order to give the reader unfamiliar with this subject a feeling for the nature of the problems that can be solved using the theory of optimal stopping rules, we now give a statement of the problem of "disruption" and the problem of the choice of the best object (also called the "secretary problem") considered in the second chapter (§ 2.6).

Let $\theta = \theta(\omega)$ be a random variable taking on the values $0, 1, \cdots$. We assume that up to the instant of time θ (we call it the time of appearance of the disruption) the observations $\xi_1, \cdots, \xi_{\theta-1}$ form a sequence of independent identically distributed random variables with distribution function $F(x)$. The observations $\xi_\theta, \xi_{\theta+1}$ are also independent identically distributed, but with distribution function $F_1(x) \neq F_0(x)$. The problem is how to use the results of the observations $\xi_1, \xi_2, \cdots$ to resolve the question of at what (Markov) time τ to declare that a "disruption" has arisen, so as to minimize the delay time $\mathbf{E}(\tau - \theta \mid \tau \geqq \theta)$ given the probability of a "false alarm" $\alpha = \mathbf{P}(\tau < \theta)$.

The solution of this problem is dealt with in the third and fourth sections of the last chapter.

---

[4] *Translators' note.* In the recent book by Chow, Siegmund and Robbins, *Great expectations*, this is referred to as a finite optimal stopping time.

The problem of choosing the best object is stated as follows. We have n objects ordered in some fashion. It is assumed that the objects arrive in random order and it is possible to determine which of them is best by pairwise comparison. We ask which object should be chosen so as to maximize the probability of choosing the best object. It is assumed here that we cannot return to a rejected object.

3. This book is based on lectures given by the author at the Mechanics-Mathematics Faculty of Moscow State University in 1966–1968 and (to a lesser extent) the Second All-Union School on Optimal Control in Shemakha (1967). In these lectures the author did not come close to covering all the problems of statistical sequential analysis, but restricted the exposition to the theory of optimal stopping rules and some of their applications. This is reflected particularly in the book's subtitle, *Optimal stopping rules.*

In the remarks at the end of the book, the sources of the results are indicated, and also literature references are given to certain works related to the material presented.

In conclusion I use this opportunity to express my deep gratitude to A.N. Kolmogorov, who introduced me to the problems of sequential analysis and whose advice I had the opportunity to follow. I would like to thank N.N. Moiseev for initiating the writing of this book. The problems of sequential analysis were the subject of many discussions with B.I. Grigelionis which were very valuable to me. I offer him my thanks. Finally I am grateful to the editor, O.V. Viškov, whose criticism helped to eliminate errors, and to M.P. Eršov, I.L. Legostaev and L.G. Straut for helping me with the preparation of the manuscript.[5)]

LIST OF BASIC PROBABILITY-THEORETIC NOTATION

$(\Omega, \mathscr{F})$ measurable space

$\mathbf{P}, \mathbf{P}_x$ probability measures (probabilities)

$(\Omega, \mathscr{F}, \mathbf{P})$ probability space

$(E, \mathscr{B})$ phase space

$\xi, \eta, \cdots$ random variables

τ, σ Markov times

$\mathbf{E}\,\xi$ mathematical expectation of ξ

$\mathbf{E}(\xi \mid \mathscr{G})$ conditional mathematical expectation of ξ with respect to the σ-algebra $\mathscr{G}$

$\mathscr{F}_t$ σ-subalgebra of $\mathscr{F}$

$\boldsymbol{T} = [0,\infty)$, $\bar{\boldsymbol{T}} = [0,\infty]$

$\boldsymbol{N} = \{0,1,\cdots\}$, $\bar{\boldsymbol{N}} = \{0,1,\cdots,\infty\}$

$X = \{\xi_t(\omega)\}$, $t \in \boldsymbol{T}$ $(t \in \boldsymbol{N})$ continuous (discrete) time random process

$X = (x_t, \mathscr{F}_t, \mathbf{P}_x)$, $x \in E$, $t \in \boldsymbol{T}$ $(t \in \boldsymbol{N})$ continuous (discrete) time Markov process

$a \wedge b = \min(a,b)$, $a \vee b = \max(a,b)$

$a^- = -\min(a,0)$, $a^+ = \max(a,0)$

(a) $\Rightarrow$ (b) (b) follows from (a)

---

5) *Translation editor's note.* This translation incorporates corrections and addenda furnished by the author especially for this edition.

CHAPTER I

MARKOV TIMES AND RANDOM PROCESSES

§ 1. Information needed from probability theory

1. Let $(\Omega,\mathscr{F})$ be a *measurable space,* i.e. a set Ω of points ω with a system $\mathscr{F}$ of subsets chosen from it which form a σ-algebra.[1]

According to the Kolmogorov axioms, all probability considerations are based on the assumption of being given a *probability space* $(\Omega,\mathscr{F},\mathbf{P})$, where $(\Omega,\mathscr{F})$ is a measurable space and $\mathbf{P}$ is a *probability measure* (probability) defined on sets from $\mathscr{F}$ and having the following properties:

$\mathbf{P}(A) \geqq 0,\quad A \in \mathscr{F}$ (nonnegativeness);

$\mathbf{P}(\Omega) = 1$ (normalizedness);

$\mathbf{P}(\bigcup_{i=1}^{\infty} A_i) = \sum_{i=1}^{\infty} \mathbf{P}(A_i)$ (countable or σ-additivity), where $A_i \in \mathscr{F}$, $A_i \cap A_j = \phi$ for $i \neq j$, and ϕ is the empty set.

The system of sets $\mathscr{F}^{\mathbf{P}}$ is called the *completion* of $\mathscr{F}$ in the measure $\mathbf{P}$ if $\mathscr{F}^{\mathbf{P}}$ contains all those sets $A \subseteq \Omega$ for which there exist $A_1, A_2 \in \mathscr{F}$ such that $A_1 \subseteq A \subseteq A_2$ and $\mathbf{P}(A_2\backslash A_1) = 0$. The system of sets $\mathscr{F}^{\mathbf{P}}$ is a σ-algebra and the measure $\mathbf{P}$ can be uniquely extended to sets from $\mathscr{F}^{\mathbf{P}}$.

The probability space $(\Omega,\mathscr{F},\mathbf{P})$ is said to be *complete* if $\mathscr{F}^{\mathbf{P}}$ coincides with $\mathscr{F}$.

Suppose $(\Omega,\mathscr{F})$ is a measurable space. We let $\bar{\mathscr{F}} = \bigcap_{\mathbf{P}} \mathscr{F}^{\mathbf{P}}$, where the intersection is taken over the system of all probability measures $\mathbf{P}$ on $(\Omega,\mathscr{F})$. The system $\bar{\mathscr{F}}$ is a σ-algebra and its sets are called *universally measurable sets* of the space $(\Omega,\mathscr{F})$.

Suppose $(\Omega,\mathscr{F})$ and $(E,\mathscr{B})$ are two measurable spaces. The function $\xi = \xi(\omega)$, defined on $(\Omega,\mathscr{F})$ and taking on values in E, is said to be $\mathscr{F}/\mathscr{B}$-measurable if for every $S \in \mathscr{B}$ the set $\{\omega: \xi(\omega) \in S\} \in \mathscr{F}$.[2] In probability theory such functions are called *random variables with values in E.*

If the measurable space $(E,\mathscr{B})$ is such that the σ-algebra $\mathscr{B}$ contains all subsets of E consisting of one point, it is called a *phase space.*

If $E = R$ is the real line and $\mathscr{B}$ is the σ-algebra of its Borel subsets, then the $\mathscr{F}/\mathscr{B}$-measurable function $\xi = \xi(\omega)$ is called simply a (real) *random variable*. In this

---

[1] A system of subsets $\mathscr{F}$ of a space Ω is called a σ-algebra if whenever $\mathscr{F}$ includes a set A, it includes its complement $\bar{A} = \Omega\backslash A$ and whenever $\mathscr{F}$ includes the sequence of sets $A_1, A_2,\ldots$ it includes their union $\bigcup_{i=1}^{\infty} A_i$ and intersection $\bigcap_{i=1}^{\infty} A_i$.

[2] The set $\{\omega: \xi(\omega) \in S\}$ is often denoted simply by $\{\xi(\omega) \in S\}$ or $(\xi(\omega) \in S)$, or even $(\xi \in S)$.

special case $\mathscr{F}/\mathscr{B}$-measurable functions are briefly said to be $\mathscr{F}$-measurable.

Suppose $\xi(\omega)$ is a nonnegative random variable. Its *mathematical expectation* (denoted by $\mathbf{E}\xi$) is by definition the Lebesgue integral[3] $\int_\Omega \xi(\omega)\mathbf{P}(d\omega)$ which (on account of the assumption $\xi(\omega) \geqq 0$) is defined, possibly taking on the value $+\infty$.

The mathematical expectation of an arbitrary real random variable $\xi(\omega)$ (also denoted by $\mathbf{E}\xi = \int_\Omega \xi(\omega)\,\mathbf{P}(d\omega)$) is defined only when one of the mathematical expectations $\mathbf{E}\xi^+$ or $\mathbf{E}\xi^-$ is finite (here $\xi^+ = \max(\xi,0)$, $\xi^- = -\min(\xi,0)$) and is set equal to $\mathbf{E}\xi^+ - \mathbf{E}\xi^-$.

The random variable $\xi(\omega)$ is said to be *integrable* if

$$\mathbf{E}\,|\,\xi\,| = \mathbf{E}\xi^+ + \mathbf{E}\xi^- < \infty.$$

The Lebesgue integral $\int_\Lambda \xi(\omega)\mathbf{P}(d\omega)$ (if it is defined, i.e. if one of the two integrals $\int_\Lambda \xi^+(\omega)\mathbf{P}(d\omega)$ or $\int_\Lambda \xi^-(\omega)\mathbf{P}(d\omega)$ is finite) on the set $\Lambda \in \mathscr{F}$ will also be denoted by $\mathbf{E}(\xi;\Lambda)$. The same is true for $\mathbf{E}\xi = \mathbf{E}(\xi,\Omega)$.

If $\mathscr{G}$ is a σ-subalgebra of $\mathscr{F}$, $\mathscr{G} \subseteq \mathscr{F}$, and $\xi = \xi(\omega)$ is a real random variable whose mathematical expectation $\mathbf{E}\xi$ is defied, then $\mathbf{E}(\xi\,|\,\mathscr{G})$ denotes the *conditional mathematical expectation of* ξ relative to $\mathscr{G}$, i.e. any $\mathscr{G}$-measurable function $\eta = \eta(\omega)$ for which $\mathbf{E}\eta$ is defined and for all $\Lambda \in \mathscr{G}$

$$\int_\Lambda \xi(\omega)\mathbf{P}(d\omega) = \int_\Lambda \eta(\omega)\mathbf{P}(d\omega). \tag{1.1}$$

On the basis of the Radon-Nikodým theorem, such a random variable $\eta(\omega)$ always exists.

If $\xi(\omega) = \chi_A(\omega)$ is the *indicator function* of the set A (in other words, the *set characteristic function of* A), then $\mathbf{E}(\chi_A(\omega)\,|\,\mathscr{G})$ is denoted by $\mathbf{P}(A\,|\,\mathscr{G})$ and is called the *conditional probability* of the set A relative to $\mathscr{G}$. Both $\mathbf{E}(\xi|\mathscr{G})$ and $\mathbf{P}(A\,|\,\mathscr{G})$ are uniquely determined by (1.1) to within sets of $\mathbf{P}$-measure zero. In other words, if $f(\omega)$ is a $\mathscr{G}$-measurable function also satisfying (1.1), then $\mathbf{E}(\xi\,|\,\mathscr{G}) = f(\omega)$ with probability one or almost surely ($\mathbf{P}$-a.s.).

If $\mathscr{A}$ is a system of subsets of the space Ω, then $\sigma(\mathscr{A})$ denotes the σ-algebra generated by the system $\mathscr{A}$, i.e. the smallest σ-algebra containing $\mathscr{A}$.

2. Let $\boldsymbol{T} = [0,\infty)$, $\bar{\boldsymbol{T}} = \boldsymbol{T} \cup \{\infty\}$, $\boldsymbol{N} = \{0,1,\ldots\}$ and $\bar{\boldsymbol{N}} = \boldsymbol{N} \cup \{\infty\}$. A family of $\mathscr{F}/\mathscr{B}$-measurable random variables $X = \{\xi_t(\omega)\}$, $t \in \boldsymbol{T}$ ($t \in \boldsymbol{N}$), is called a continuous (discrete) time *random process* with values in $(E,\mathscr{B})$. A discrete time random process is also called a *random sequence*.

For fixed $\omega \in \Omega$ the function of time $\xi_t(\omega)$, $t \in \boldsymbol{T}$ (or $t \in \boldsymbol{N}$), is called the *trajectory* corresponding to the *elementary event* ω. For clarity it is sometimes convenient to say that $\xi_t(\omega)$, $t \in \boldsymbol{T}$ (or $t \in \boldsymbol{N}$), is the trajectory of motion of a particle (or system).

If $\mathscr{A}_t$ is the algebra of subsets of Ω generated by the sets $\{\omega\colon \xi_s(\omega) \in \Gamma\}$, $\Gamma \in \mathscr{B}$,

---

[3] The notation $\int_\Omega \xi(\omega)d\mathbf{P}$ is also often used for the Lebesgue integral.

$s \leqq t$, then we let $\sigma\{\omega: \xi_s(\omega), s \leqq t\} = \sigma(\mathscr{A}_t)$, $\mathbf{E}(\eta \mid \xi_s, s \leqq t) = \mathbf{E}(\eta \mid \sigma(\mathscr{A}_t))$ and $\mathbf{P}(A \mid \xi_s, s \leqq t) = \mathbf{P}(A \mid \sigma(\mathscr{A}_t))$, where η is an $\mathscr{F}$-measurable random variable for which the mathematical expectation $\mathbf{E}\eta$ is defined, and $A \in \mathscr{F}$. Sometimes we also use the notation $\mathbf{E}(\eta \mid \xi_0^t) = \mathbf{E}(\eta \mid \sigma(\mathscr{A}_t))$ and $\mathbf{P}(A\xi_0^t) = \mathbf{P}(A \mid \sigma(A_t))$.

The random process $X = \{\xi_t(\omega)\}$, $t \in \boldsymbol{T}$, is said to be *measurable* if for all $S \in \mathscr{B}$

$$\{(t,\omega): \xi_t(\omega) \in S\} \in \mathscr{F} \times \mathscr{B}(\boldsymbol{T}),$$

where $\mathscr{B}(\boldsymbol{T})$ is the σ-algebra of Borel sets on $\boldsymbol{T} = [0,\infty)$.

The random process X is said to be *adapted to* the family of σ-algebras $F = \{\mathscr{F}_t\}$, $t \in \boldsymbol{T}$, if $\{\omega: \xi_t(\omega) \in S\} \in \mathscr{F}_t$ for all $t \in \boldsymbol{T}$ and $S \in \mathscr{B}$.

We say that the process X is *progressively measurable* (relative to $F = \{\mathscr{F}_t\}$, $t \in \boldsymbol{T}$)[4] if for each $t \in \boldsymbol{T}$

$$\{(\omega, s): \xi_s(\omega) \in S, s \leqq t\} \in \mathscr{F}_t \times \mathscr{B}([0, t]),$$

where $\mathscr{B}([0, t])$ is the σ-algebra of Borel sets in $[0, t]$.

Every progressively measurable (relative to $F = \{\mathscr{F}_t\}$) process X is measurable and adapted to F. The reverse assertion also holds: if the process X is measurable and adapted to $F = \{\mathscr{F}_t\}$, then it is progressively measurable relative to $F = \{\mathscr{F}_t\}$ (more precisely, there exists a progressively measurable process $X' = \{\xi'_t(\omega)\}$, $t \in \boldsymbol{T}$, equivalent to X.[5]) It is known that every process which is adapted to $F = \{\mathscr{F}_t\}$ and has right (left) continuous trajectories is progressively measurable **[49]**.

The results set forth in the above section also remain valid in the case where E is a locally compact Hausdorff space with a countable basis **[49]**.

§ 2. Markov times

1. In this section we give definitions and present the basic properties of Markov times, which will be used later in the solution of various problems on optimal stopping of Markov processes. The entire presentation will be made for the case of continuous time. The corresponding definitions and results carry over automatically to the case of discrete time and, as a rule, become simpler.

Let $(\Omega,\mathscr{F})$ be a measurable space, let $\boldsymbol{T} = [0,\infty)$, and let $F = \{\mathscr{F}_t\}$, $t \in \boldsymbol{T}$, be a nondecreasing sequence of σ-subalgebras of $\mathscr{F}$: $\mathscr{F}_s \subseteq \mathscr{F}$, $\mathscr{F}_s \subseteq \mathscr{F}_t$, $s \leqq t$.

DEFINITION 1. A random variable (i.e. an $\mathscr{F}$-measurable function) $\tau = \tau(\omega)$ with values in $\overline{\boldsymbol{T}} = [0, \infty]$ is called a *Markov time* (relative to the sequence $F = \{\mathscr{F}_t\}$)[6] if $\{\omega: \tau \leqq t\} \in \mathscr{F}_t$ for each $t < \infty$.

Markov times (M.t.'s) are random variables not depending on the future.

DEFINITION 2. If $\tau = \tau(\omega)$ is a M.t. (relative to $F = \{\mathscr{F}_t\}$), then $\mathscr{F}_\tau$ denotes the

---

4) In those cases which do not give rise to ambiguity, we omit the $t \in \boldsymbol{T}$.

5) The processes $X = \{\xi_t(\omega)\}$ and $X' = \{\xi'_t(\omega)\}$, $t \in \boldsymbol{T}$, are said to be *equivalent* if $\mathbf{P}\{\xi_t(\omega) \neq \xi'_t(\omega)\} = 0$ for all $t \in \boldsymbol{T}$.

6) In cases where no ambiguity arises, the words "relative to the sequence $F = \{\mathscr{F}_t\}$" will be omitted.

collection of those sets $A \in \{\omega: \tau < \infty\}$ for which $A \cap \{\tau \leqq t\} \in \mathscr{F}_t$ for all $t \in \boldsymbol{T}$.

It is easy to see that $\mathscr{F}_\tau$ is a σ-algebra, and if $\tau(\omega) = s$ for all $\omega \in \Omega$, then $\mathscr{F}_\tau$ coincides with $\mathscr{F}_s$. The intuitive meaning of the σ-algebra $\mathscr{F}^\tau$ is the following. By $\mathscr{F}_t$ we mean the collection of events connected with some physical process and observable up to time t. Then $\mathscr{F}_\tau$ is the collection of events observable over the random time τ.

2. For each $t \in \boldsymbol{T}$ we set $\mathscr{F}_{t+0} = \bigcap_{s>t} \mathscr{F}_s$ and $\mathscr{F}_\infty = \sigma(\bigcup_{s<\infty} \mathscr{F}_s)$.

DEFINITION 3. The sequence $F = \{\mathscr{F}_t\}$ is said to be *right continuous* if $\mathscr{F}_t = \mathscr{F}_{t+0}$ for all $t \in \boldsymbol{T}$.

LEMMA 1. *If $\{\tau \leqq t\} \in \mathscr{F}_t$, for each $t \in \boldsymbol{T}$, then for each $t \in \boldsymbol{T}$*

$$\{\tau < t\} \in \mathscr{F}_t \tag{1.2}$$

and consequently $\{\tau = t\} \in \mathscr{F}_t$.

PROOF. Since $\{\tau < t\} = \bigcup_{k=1}^{\infty} \{\tau \leqq t - (1/k)\}$ and $\mathscr{F}_{t-(1/k)} \subseteq \mathscr{F}_t$ we have $\{\tau < t\} \in \mathscr{F}_t$.

The assertion converse to (1.2) is not true in general. However, we do have

LEMMA 2. *If the family $\{\mathscr{F}_t\}$ is right continuous and $\{\tau < t\} \in \mathscr{F}_t$ for each $t \in \boldsymbol{T}$, then $\{\tau \leqq t\} \in \mathscr{F}_t$ for each $t \in \boldsymbol{T}$.*

PROOF. If $\{\tau < t\} \in \mathscr{F}_t$, then $\{\tau \leqq t\} \in \mathscr{F}_{t+\varepsilon}$ for each $\varepsilon > 0$. Consequently $\{\tau \leqq t\} \in \mathscr{F}_{t+0} = \mathscr{F}_t$.

From this lemma it follows that in the case of right continuous families $\{\mathscr{F}_t\}$ it suffices to establish that $\{\tau < t\} \in \mathscr{F}_t$, $t \in \boldsymbol{T}$, in order to verify that the random variable $\tau = \tau(\omega)$ is a M.t.

In the general case the condition $\{\tau < t\} \in \mathscr{F}_t$, $t \in \boldsymbol{T}$, is weaker than the condition $\{\tau \leqq t\} \in \mathscr{F}_t$, $t \in \boldsymbol{T}$. To see this we set $\Omega = \boldsymbol{T}$, let $\mathscr{F}$ be the σ-algebra of Lebesgue measurable sets in $\boldsymbol{T}$, and set $x_t(\omega) = 0$ for $t \leqq \omega$, $x_t(\omega) = 1$ for $t > \omega$, and $\mathscr{F}_t = \sigma\{\omega: x_s(\omega), s \leqq t\}$. Then the random variable $\tau = \inf\{t \geqq 0: x_t(\omega) = 1\}$ satisfies the condition $\{\tau < t\} \in \mathscr{F}_t$, while at the same time $\{\tau \leqq t\} \notin \mathscr{F}_t$, $t \in \boldsymbol{T}$.

Suppose $X = \{\xi_t(\omega)\}$, $t \in \boldsymbol{T}$, is a real random process given on some probability space $(\Omega, \mathscr{F}, \mathbf{P})$. The most important example of a nondecreasing sequence of σ-algebras $\{\mathscr{F}_t\}$ is the sequence $\mathscr{F}_t = \mathscr{F}_t^\xi = \sigma\{\omega: \xi_s(\omega), s \leqq t\}$.

The next theorem turns out to be valuable in determining whether a random variable $\tau = \tau(\omega)$ (with values in $\bar{\boldsymbol{T}}$) is Markov relative to the family $\{\mathscr{F}_t^\xi\}$.

THEOREM 1. *Let $\Omega = \{\omega\}$ be the space of all real functions $f(s)$, $s \geqq 0$. In order for the random variable $\tau = \tau(\omega)$ with values in $\bar{\boldsymbol{T}}$ to be a Markov time relative to $\{\mathscr{F}_t^\xi\}$, it is necessary and sufficient for it to be measurable with respect to the σ-algebra $\mathscr{F}_\infty^\xi = \sigma(\bigcup_{t \in \boldsymbol{T}} \mathscr{F}_t^\xi)$ and that $\tau(\omega') = \tau(\omega)$ for all $t < \infty$, $\omega \in \Omega$, $\omega' \in \Omega$ such that*

$$\tau(\omega) \leqq t, \quad \xi_s(\omega) = \xi_s(\omega'), \quad s \leqq t.$$

The proof of this theorem is based on Lemma 3; in order to state the latter we need to introduce some new concepts and notation.

We say that the points ω and ω' in Ω are t-equivalent ($\omega' \overset{t}{\sim} \omega$) if $\xi_s(\omega) = \xi_s(\omega')$ for all $s \leqq t$.

We also say that the set A belongs to the system of sets $\tilde{\mathscr{F}}_t$ if $A \in \mathscr{F}^\xi_\infty$ and A is completed by t-equivalent points (i.e. $\omega \in A$ and $\omega \overset{t}{\sim} \omega' \Rightarrow \omega' \in A$). It can be verified that for every t the system $\tilde{\mathscr{F}}_t$ forms a σ-algebra.

LEMMA 3. *The σ-algebras $\tilde{\mathscr{F}}^\xi_t$ and $\mathscr{F}^\xi_t$ coincide.*

PROOF. Let α_t, $t \in \boldsymbol{T}$, denote the mapping (if, of course, it exists) of Ω into Ω such that

$$\xi_s(\alpha_t\omega) = \xi_{s\wedge t}(\omega),$$

where $s \wedge t = \min(s,t)$. In particular, if the space Ω of elementary events is the space of all real functions $f(s)$, $s \geqq 0$, then the mapping α_t exists and takes the function $f(s)$, $s \geqq 0$, into the function

$$(\alpha_t f)(s) = \begin{cases} f(s), & s \leqq t, \\ f(t), & s \geqq t. \end{cases}$$

We note that for each $t \in \boldsymbol{T}$, α_t is a measurable mapping of $(\Omega, \mathscr{F}^\xi_t)$ into $(\Omega, \mathscr{F}^\xi_\infty)$. Consequently, if $A \in \mathscr{F}^\xi_\infty$, then $\alpha_t^{-1}(A) \in \mathscr{F}^\xi_t$. We assume that the set A is completed by t-equivalent points. Then, obviously, $A \in \tilde{\mathscr{F}}^\xi_t$. We shall show that in this case $A \in \mathscr{F}^\xi_t$. Indeed, suppose $\omega \in A$. Then the point $\alpha_t\omega$ also belongs to the set A and hence $\alpha_t^{-1}(A) \supseteq A$. In fact $\alpha_t^{-1}(A) = A$, since if $\omega \in \alpha_t^{-1}(A)$ then $\alpha_t\omega \in A$, and since $\omega \overset{t}{\sim} \alpha_t\omega$ we have $\omega \in A$. But $\alpha_t^{-1}(A) \in \mathscr{F}^\xi_t$, and therefore $A \in \mathscr{F}^\xi_t$.

It suffices to carry out the proof of the converse relationship ($A \in \mathscr{F}^\xi_t \Rightarrow A \in \tilde{\mathscr{F}}^\xi_t$) for sets A of the form $A = \{\omega : \xi_s(\omega) \in [a,b]\}$, $s \leqq t$. If $\omega \in A$ and $\omega' \overset{t}{\sim} \omega$, then, in view of the equality $\xi_s(\omega) = \xi_s(\omega')$, $s \leqq t$, we obviously have $\omega' \in A$. Therefore the set $A \in \mathscr{F}^\xi_t \subseteq \mathscr{F}^\xi_\infty$ is completed by t-equivalent points, and hence $A \in \tilde{\mathscr{F}}^\xi_t$.

We proceed with the proof of Theorem 1.

NECESSITY. Suppose $\tau = \tau(\omega)$ is a M.t., and for $t \in \boldsymbol{T}$, $\omega \in \Omega$ and $\omega' \in \Omega$ let

$$\tau(\omega) \leqq t, \quad \omega' \overset{t}{\sim} \omega.$$

For given ω we let $u = \tau(\omega)$. Then

$$\omega \in A = \{\tau(\omega) = u\} \in \mathscr{F}_u.$$

By Lemma 3, $\mathscr{F}^\xi_u$ coincides with $\tilde{\mathscr{F}}^\xi_u$. Therefore the set $A \in \tilde{\mathscr{F}}^\xi_u$, and since $\omega' \overset{u}{\sim} \omega$, then $\omega' \in A$. Thus $\tau(\omega') = u$ and hence $\tau(\omega) = \tau(\omega')$.

SUFFICIENCY. Suppose $\tau = \tau(\omega)$ is an $\mathscr{F}_\infty^\xi$-measurable random variable and for all $t \in \boldsymbol{T}$, $\omega \in \Omega$ and $\omega' \in \Omega$ such that $\tau(\omega) \leqq t$ and $\omega' \overset{t}{\sim} \omega$, we have $\tau(\omega') = \tau(\omega)$. It is clear that the set $A = \{\tau(\omega) \leqq t\} \in \mathscr{F}_\infty^\xi$ contains all t-equivalent points, i.e. $A \in \tilde{\mathscr{F}}_t^\xi$. Hence $A \in \mathscr{F}_t^\xi$ by Lemma 3.

From this theorem it follows that in the case of a sequence $\{\mathscr{F}_t^\xi\}$ we can make use of another definition of M.t., equivalent to the one given earlier.

DEFINITION 4. An $\mathscr{F}_\infty^\xi$-measurable random variable $\tau = \tau(\omega)$ with values in $\bar{\boldsymbol{T}} = [0,\infty]$ is said to be a M.t. if for all $t \in \boldsymbol{T}$, $\omega \in \Omega$ and $\omega' \in \Omega$ such that $\tau(\omega) \leqq t$ and $\xi_s(\omega) = \xi_s(\omega')$, $s \leqq t$, we have $\tau(\omega) = \tau(\omega')$.

We note that in work on sequential analysis, M.t.'s are usually taken to mean random variables $\tau = \tau(\omega)$ satisfying Definition 4, which, as we can see, is indeed equivalent to Definition 1.

REMARK. Theorem 1 allows the following equivalent formulation.

THEOREM 1*. *Let $\Omega = \{\omega\}$ be the space of all real functions $f(s)$, $s \geqq 0$. In order for the random variable $\tau = \tau(\omega)$ with values in $\bar{\boldsymbol{T}}$ to be a Markov time relative to $\{\mathscr{F}_t^\xi\}$, it is necessary and sufficient for it to be measurable with respect to $\mathscr{F}_\infty^\xi$ and that the equality $\tau(\omega') = t$ hold for all $t < \infty$ and $\omega, \omega' \in \Omega$ such that $\tau(\omega) = t$ and $\xi_s(\omega) = \xi_s(\omega')$, $s \leqq t$.*

From this theorem and Lemma 3 we have the following corollaries.

COROLLARY 1. *The random variable $\tau = \tau(\omega)$ is a Markov time relative to $\{\mathscr{F}_t^\xi\}$ if and only if*

(1) *τ is an $\mathscr{F}_\infty^\xi$-measurable function,*

(2) *$\{\tau = t\} \in \mathscr{F}_t^\xi$ for all $t \geqq 0$.*

The necessity of these conditions is obvious. On the other hand, if these conditions are satisfied, then $\{\tau \leqq t\} \in \tilde{\mathscr{F}}_t^\xi = \mathscr{F}_t^\xi$.

COROLLARY 2. *The set $A \subseteq \{\tau < \infty\}$ belongs to $\mathscr{F}_\tau^\xi$ if and only if*

(1) *$A \in \mathscr{F}_\infty^\xi$,*

(2) *$A \cap \{\tau = t\} \in \mathscr{F}_t^\xi$ for all $t \geqq 0$.*

The necessity of the conditions is clear. For the sufficiency, we need to establish that $A \cap \{\tau \leqq t\} \in \mathscr{F}_t^\xi$ for all $t \geqq 0$. We have $A \cap \{\tau \leqq t\} \in \mathscr{F}_\infty^\xi$. Further, the set $A \cap \{\tau \leqq t\}$ is completed by t-equivalent points and, consequently, $A \cap \{\tau \leqq t\} \in \tilde{\mathscr{F}}_t^\xi = \mathscr{F}_t^\xi$.

COROLLARY 3. *If the function $\xi_\tau(\omega)$ is $\mathscr{F}_\infty^\xi$-measurable, then $\xi_\tau(\omega)$ is already $\mathscr{F}_\tau^\xi$-measurable. Indeed, $\{\xi_\tau(\omega) \in \Gamma\} \in \mathscr{F}_\infty^\xi$ and $\{\xi_\tau \in \Gamma\} \cap \{\tau = t\} = \{\xi_t \in \Gamma\} \cap \{\tau = t\} \in \mathscr{F}_t^\xi$. By Corollary 2, $\{\xi_\tau \in \Gamma\} \in \mathscr{F}_\tau^\xi$.*

3. The definitions given above relate to the case of continuous time $t \in \boldsymbol{T}$. As already mentioned, all the concepts and results presented above also hold in the case of an arbitrary set of values t as long as they are ordered. In particular, we assume that the time t is discrete:[7] $t \in \boldsymbol{N} = \{0,1,\cdots\}$.

LEMMA 4. *Suppose $n \in \boldsymbol{N}$ and $\tau = \tau(\omega)$ takes on values in $\bar{\boldsymbol{N}} = \boldsymbol{N} \cup \{\infty\}$. The conditions $\{\tau \leqq n\} \in \mathscr{F}_n$ and $\{\tau = n\} \in \mathscr{F}_n$, $n \in \boldsymbol{N}$, are equivalent.*

PROOF. As in Lemma 1, $\{\tau \leqq n\} \in \mathscr{F}_n \Rightarrow \{\tau = n\} \in \mathscr{F}_n$. The converse assertion follows from the fact that $\{\tau \leqq n\} = \bigcup_{k \leqq n} \{\tau = k\} \in \mathscr{F}_n$.

4. LEMMA 5. *If τ_1 and τ_2 are M.t.'s, then $\tau_1 \wedge \tau_2 = \min(\tau_1,\tau_2)$, $\tau_1 \vee \tau_2 = \max(\tau_1,\tau_2)$ and $\tau_1 + \tau_2$ are also Markov times.*

Let $\{\tau_n\}$, $n = 1,2,\cdots$, be a sequence of M.t.'s. Then $\sup_n \tau_n$ is also a M.t. If, moreover, the sequence $\{\mathscr{F}_t\}$ is right continuous, then $\inf_n \tau_n$, $\limsup_n \tau_n$ and $\liminf_n \tau_n$ are also Markov times.

The proof of the first three assertions follows from the relationships where

$$\begin{aligned} \{\tau_1 \wedge \tau_2 \leqq t\} &= \{\tau_1 \leqq t\} \cup \{\tau_2 \leqq t\}, \\ \{\tau_1 \vee \tau_2 \leqq t\} &= \{\tau_1 \leqq t\} \cap \{\tau_2 \leqq t\}, \\ \{\tau_1 + \tau_2 \leqq t\} &= \{\tau_1 = 0, \tau_2 = t\} \\ \cup \{\tau_1 = t, \tau_2 = 0\} &\bigcup_{\substack{a+b<t \\ a,b \geqq 0}} \{\tau_1 < a\} \cap \{\tau_2 < b\}, \end{aligned}$$

where a and b are rational numbers.

The proof of the remaining assertions is based on the facts that

$$\begin{aligned} \left\{\sup_n \tau_n \leqq t\right\} &= \bigcap_n \{\tau_n \leqq t\} \in \mathscr{F}_t, \\ \left\{\inf_n \tau_n < t\right\} &= \bigcup_n \{\tau_n < t\} \in \mathscr{F}_t \end{aligned}$$

and that

$$\begin{aligned} \left\{\limsup_n \tau_n < t\right\} &= \bigcup_{k=1}^{\infty} \bigcup_{n=1}^{\infty} \bigcap_{m=n}^{\infty} \left\{\tau_n < t - \frac{1}{k}\right\}, \\ \left\{\liminf_n \tau_n > t\right\} &= \bigcup_{k=1}^{\infty} \bigcap_{n=1}^{\infty} \bigcup_{m=n}^{\infty} \left\{\tau_n > t + \frac{1}{k}\right\} \end{aligned}$$

for

$$\limsup_n \tau_n = \inf_{n \geqq 1} \sup_{m \geqq n} \tau_m, \qquad \liminf_n \tau_n = \sup_{n \geqq 1} \inf_{m \geqq n} \tau_m$$

LEMMA 6. *Every Markov time $\tau = \tau(\omega)$ (relative to $\{\mathscr{F}_t\}$) is an $\mathscr{F}_\tau$-measurable*

[7] In this case the concepts of σ-algebras $\mathscr{F}_{t+0}$ and right continuity of the family $\{\mathscr{F}_t\}$ are meaningless.

random variable. If $\tau(\omega)$ and $\sigma(\omega)$ are to M.t.'s and $\tau(\omega) \leqq \sigma(\omega)$, then $\mathscr{F}_\tau \subseteq \mathscr{F}_\sigma$.

PROOF. Let $A = \{\tau \leqq s\}$. We have to show that $A \cap \{\tau \leqq t\} \in \mathscr{F}_t$ for all $t \in \boldsymbol{T}$. But

$$\{\tau \leqq s\} \cap \{\tau \leqq t\} = \{\tau \leqq t \wedge s\} \in \mathscr{F}_{t\wedge s} \subseteq \mathscr{F}_t,$$

and hence the M.t. τ is an $\mathscr{F}_\tau$-measurable random variable.

Now suppose that $A \in \mathscr{F}_\tau$. Then

$$A \cap \{\sigma \leqq t\} = (A \cap \{\tau \leqq t\}) \cap \{\sigma \leqq t\} \in \mathscr{F}_t$$

and consequently $A \in \mathscr{F}_\sigma$.

LEMMA 7. *Let $\{\tau_n\}$ be a sequence of M.t.'s relative to the right continuous system of σ-algebras $\{\mathscr{F}_t\}$, and suppose $\tau = \inf_n \tau_n$. Then $\mathscr{F}_\tau = \bigcap_n \mathscr{F}_{\tau_n}$.*

PROOF. On the strength of Lemma 5, τ is a M.t. Therefore, according to Lemma 6, $\mathscr{F}_\tau \subseteq \bigcap_n \mathscr{F}_{\tau_n}$. On the other hand, if $A \in \bigcap_n \mathscr{F}_{\tau_n}$, then

$$A \cap \{\tau < t\} = A \cap \Big(\bigcup_n \{\tau_n < t\}\Big) = \bigcup_n (A \cap \{\tau_n < t\}) \in \mathscr{F}_t,$$

whence, by the right continuity of $\{\mathscr{F}_t\}$, it is easy to obtain that $A \in \mathscr{F}_\tau$.

LEMMA 8, *Let τ and σ be two M.t.'s relative to $\{\mathscr{F}_t\}$. Then each of the events $\{\tau < \sigma\}$, $\{\tau \leqq \sigma\}$, $\{\tau \geqq \sigma\}$ and $\{\tau = \sigma\}$ belongs to $\mathscr{F}_\tau$ and $\mathscr{F}_\sigma$.*

PROOF. For each $t \in \boldsymbol{T}$

$$\{\tau < \sigma\} \cap \{\sigma \leqq t\} = \bigcup_{r<t} (\{\tau < r\} \cap \{r < \sigma \leqq t\}) \in \mathscr{F}_t,$$

where r are rational numbers, whence $\{\tau < \sigma\} \in \mathscr{F}_\sigma$.

On the other hand,

$$\{\tau < \sigma\} \cap \{\tau \leqq t\} = \bigcup_{r<t} (\{\tau \leqq r\} \cap \{r < \sigma\}) \cup (\{\tau \leqq t\} \cap \{t < \sigma\}) \in \mathscr{F}_t,$$

i.e. $\{\tau < \sigma\} \in \mathscr{F}_\tau$.

It can be established analogously that $\{\sigma < \tau\} \in \mathscr{F}_\tau$ and $\{\sigma < \tau\} \in \mathscr{F}_\sigma$. Consequently $\{\tau \leqq \sigma\}$, $\{\sigma \leqq \tau\}$ and $\{\tau = \sigma\}$ belong to both $\mathscr{F}_\tau$ and $\mathscr{F}_\sigma$.

LEMMA 9. *If the process $X = \{\xi_t(\omega)\}$, $t \in \boldsymbol{T}$, given in the measurable space $(E,\mathscr{B})$, is progressively measurable with respect to the system $\{\mathscr{F}_t\}$ and $\tau = \tau(\omega)$ is a M.t. (relative to $\{\mathscr{F}_t\}$) such that $\mathbf{P}(\tau < \infty) = 1$, then the function $\xi_{\tau(\omega)}(\omega)$ is $\mathscr{F}_\tau/\mathscr{B}$-measurable.*

PROOF.[8] Suppose $S \in \mathscr{B}$ and $t \in \boldsymbol{T}$. We have to establish that

[8] Compare the proof of Corollary 3 to Theorem 1.

$$\{\xi_{\tau(\omega)}(\omega) \in S\} \cap \{\tau \leqq t\} \in \mathscr{F}_t.$$

Let $\sigma = \min(\tau,t)$. Then

$$\begin{aligned}\{\xi_\tau(\omega) \in S\} \cap \{\tau \leqq t\} &= \{\xi_\tau \in S\} \cap [\{\tau < t\} \cup \{\tau = t\}] \\ &= [\{\xi_\sigma(\omega) \in S\} \cap \{\sigma < t\}] \cup [\{\xi_\tau(\omega) \in S\} \cap \{\tau = t\}].\end{aligned}$$

Clearly $[\{\xi_\tau \in S\} \cap \{\tau = t\}] \in \mathscr{F}_t$. If we now show that $\xi_\sigma(\omega)$ is an $\mathscr{F}_t/\mathscr{B}$-measurable function, then $\{\xi_\sigma \in S\} \cap \{\sigma < t\} \in \mathscr{F}_t$. But, indeed, the mapping $\omega \to (\omega, \sigma(\omega))$ is a measurable mapping of $(\Omega,\mathscr{F}_t)$ into $(\Omega \times [0,t], \mathscr{F}_t \times \mathscr{B}([0,t]))$, and the mapping $(\omega,s) \to \xi_s(\omega)$ of the space $(\Omega \times [0,t], \mathscr{F}_t \times \mathscr{B}([0,t]))$ into $(E,\mathscr{B})$ is also measurable due to the progressive measurability of the process X. Consequently the mapping of $(\Omega,\mathscr{F}_t)$ into $(E,\mathscr{B})$ given by $\xi_\sigma(\omega)$ is measurable, being the result of the successive application of two measurable mappings.

5. We give several examples of Markov times.

Let $X = \{\xi_t(\omega)\}$, $t \in \boldsymbol{T}$, be a real process, and let $\mathscr{F}_t = \sigma\{\omega\colon \xi_s, s \leqq t\}$. It is obvious that the process X is adapted to the family $\{\mathscr{F}_t\}$. Let A be a Borel set on the number line and

$$\sigma_A = \inf\{t \geqq 0\colon \xi_t(\omega) \in A\}, \tag{1.3}$$
$$\tau_A = \inf\{t > 0\colon \xi_t(\omega) \in A\} \tag{1.4}$$

be respectively the so-called "first entry time of A" and the "first hitting time of A." We write $\sigma_A = \infty$ and $\tau_A = \infty$ if the sets $\{\cdot\}$ in (1.3) and (1.4) are empty.

Below, the times σ_A and τ_A (not coinciding only in the case where $\xi_0(\omega) \in A$ and there exists $\varepsilon > 0$ such that $\xi_t(\omega) \notin A$ for all $t \in (0,\varepsilon)$) will play an important role in finding optimal stopping rules. It can easily be shown that σ_A and τ_A have the following properties:

$$A \subseteq B \Rightarrow \sigma_A \geqq \sigma_B, \quad \tau_A \geqq \tau_B \tag{1.5}$$
$$\sigma_{A\cup B} = \min(\sigma_A, \sigma_B), \quad \tau_{A\cup B} = \min(\tau_A, \tau_B), \tag{1.6}$$
$$\sigma_{A\cap B} \geqq \max(\sigma_A, \sigma_B), \quad \tau_{A\cap B} \geqq \max(\tau_A, \tau_B), \tag{1.7}$$

and if $A = \bigcup_n A_n$, then

$$\sigma_A = \inf_n \sigma_{A_n} \text{ and } \tau_A = \inf_n \tau_{A_n}. \tag{1.8}$$

Lemma 10. *If the real process $X = \{\xi_t(\omega)\}$, $t \in \boldsymbol{T}$, is right continuous, $\mathscr{F}_{t+0} = \mathscr{F}_t$ and C is an open set, then σ_C and τ_C are Markov times.*

Proof. Let $D = R\backslash C$. Then, by the right continuity of the trajectories and the closedness of the set D,

$$\{\sigma_C(\omega) \geqq t\} = \{\xi_s(\omega) \in D, s < t\} = \bigcap_{r<t} \{\xi_r(\omega) \in D\},$$

where r are rational numbers. Consequently

$$\{\sigma_C(\omega) < t\} = \bigcup_{r<t} \{\xi_r(\omega) \in C\} \in \mathscr{F}_t.$$

By the assumption $\mathscr{F}_t = \mathscr{F}_{t+0}$ and Lemma 2, it now follows that $\sigma_C(\omega)$ is a M.t. The proof is carried out analogously for $\tau_C(\omega)$.

Using the method applied in the proof of Lemma 10, we can also establish that, for example, $\sigma_D = \inf\{t \geqq 0: \xi_t \in D\}$, where D is a closed set and the process $X = \{\xi_t(\omega)\}$, $t \in \boldsymbol{T}$, is continuous and is Markov relative to the system $\{\mathscr{F}_t\}$, $\mathscr{F}_t = \sigma\{\omega: \xi_s, s \leqq t\}$.

All these results on the measurability of the times σ_A and τ_A can be obtained from the following theorem.

THEOREM 2. *Suppose $X = \{\xi_t(\omega)\}$, $t \in \boldsymbol{T}$, is a progressively measurable (relative to $\{\mathscr{F}_t\}$) random process given in the measurable space $(E, \mathscr{B})$. Also suppose $\mathscr{F}_t = \mathscr{F}_t^{\mathbf{P}}$ and $\mathscr{F}_{t+0} = \mathscr{F}_t$, $t \in \boldsymbol{T}$. Then for every universally measurable set $B \in \overline{\mathscr{B}}$ ($\overline{\mathscr{B}} = \bigcap_\mu \mathscr{B}^\mu$; see § 1) the times*

$$\sigma_B^s = \inf\{t \geqq s: \xi_t \in B\}, \qquad \tau_B^s = \inf\{t > s: \xi_t \in B\},$$

where $s \geqq 0$, are Markov relative to $\{\mathscr{F}_t\}$, $t \in \boldsymbol{T}$.

The proof can be found in **[49]**, Chapter IV, Theorem 52, or in § 2 of the Addendum to **[24]**.

§ 3. Markov random processes

1. *Definitions.* We give the basic definitions and properties of discrete and continuous time Markov processes to the extent that they are needed for our investigation of optimal stopping problems.

Let $(\Omega,\mathscr{F})$ be a measurable space of elementary events $\omega \in \Omega$ and let $(E,\mathscr{B})$ be a phase space. We assume that for each $t \in \boldsymbol{Z}$ ($\boldsymbol{Z} = \boldsymbol{T} = [0,\infty)$ in the case of continuous time and $\boldsymbol{Z} = \boldsymbol{N} = \{0,1,\cdots\}$ in the case of discrete time) we identify σ-algebras $\mathscr{F}_t$ in $\mathscr{F}$ such that $\mathscr{F}_t \subseteq \mathscr{F}$ and $\mathscr{F}_t \supseteq \mathscr{F}_s$ whenever $t \geqq s$. Further suppose that $\{x_t(\omega)\}$, $t \in \boldsymbol{Z}$, $\omega \in \Omega$, is a family of random variables $x_t = x_t(\omega)$ defined on $(\Omega,\mathscr{F})$ with values in E and adapted to the system of σ-algebras $F = \{\mathscr{F}_t\}$, $t \in \boldsymbol{Z}$, and that for each $x \in E$ we are given a probability measure $\mathbf{P}_x$ on the σ-algebra $\mathscr{F}$.

DEFINITION 1. The system $X = \{x_t,\mathscr{F}_t,\mathbf{P}_x\}$, $t \in \boldsymbol{Z}$, is called a (homogeneous, nonterminating) *Markov process* with values in the phase space $(E,\mathscr{B})$ if the following conditions are satisfied:

1) $\mathbf{P}_x(A)$ is a $\mathscr{B}$-measurable function of x for each $A \in \mathscr{F}$.

2) For all $x \in E$, $B \in \mathscr{B}$ and $u, t \in \boldsymbol{Z}$

$$\mathbf{P}_x(x_{t+u}(\omega) \in B \mid \mathscr{F}_t) = \mathbf{P}_{x_t}(x_u \in B) \qquad (\mathbf{P}_x\text{-a.s.}). \tag{1.9}$$

3) $\mathbf{P}_x(x_0 = x) = 1$, $x \in E$.

4) For all $\omega \in \Omega$ and $t \in \boldsymbol{Z}$ there is $\omega' \in \Omega$ such that $x_s(\omega') = x_{s+t}(\omega)$ for all $s \in \boldsymbol{Z}$.

If $\boldsymbol{Z} = \boldsymbol{N}$, then $X = (x_t, \mathscr{F}_t, \mathbf{P}_x)$ is also called a *discrete time Markov process*, or a *Markov random sequence*.

Condition 2) expresses the Markov principle of independence of the "future" from the "past" for fixed "present." Condition 4) means that the original space Ω of elementary events must be sufficiently "rich" and that the set of trajectories $\{x_t(\omega)\}$, $t \in \boldsymbol{Z}$, has a certain homogeneity.

We let $\mathscr{F}'_t = \sigma\{\omega\colon x_s(\omega),\ s \leqq t\}$. It is easy to see that when $X = \{x_t, \mathscr{F}_t, \mathbf{P}_x\}$ is Markov, the process $X' = \{x_t, \mathscr{F}'_t, \mathbf{P}_x\}$ is also Markov and $\mathscr{F}'_t \subseteq \mathscr{F}_t$, $t \in \boldsymbol{Z}$.

It is often convenient to assume that the space of elementary events $\Omega = E^{\boldsymbol{Z}}$, i.e., it is the space of functions $\omega = \omega(t)$ defined for $t \in \boldsymbol{Z}$ with values in E. This assumption does not restrict generality, since we can construct a new Markov process with $\Omega = E^{\boldsymbol{Z}}$ which, from the point of view of finite-dimensional distributions, is equivalent to the process X ([**12**], Chapter I, Theorem 4.3).

DEFINITION 2. A progressively measurable Markov process $X = (x_t, \mathscr{F}_t, \mathbf{P}_x)$, $t \in \boldsymbol{Z}$, is said to be *strong Markov* if for each Markov time τ (relative to the system $F = \{\mathscr{F}_t\}$, $t \in \boldsymbol{Z}$) the following stronger version of condition 2) is satisfied:

2′) For all $x \in E$, $B \in \mathscr{B}$, $u \in \boldsymbol{Z}$ on the set $\{\omega\colon \tau(\omega) < \infty\}$

$$\mathbf{P}_x\big(x_{\tau+u}(\omega) \in B \,\big|\, \mathscr{F}_\tau\big) = \mathbf{P}_{x_\tau}(x_u \in B) \qquad (\mathbf{P}_x\text{-a.s.}). \tag{1.10}$$

It is known that a discrete time Markov process is always strong Markov [**24**]. In the case of continuous time this is generally not so.

DEFINITION 3. A progressively measurable Markov process $X = (x_t, \mathscr{F}_t, \mathbf{P}_x)$, $t \in \boldsymbol{T}$, is said to be quasi-left-continuous if for every nondecreasing sequence of Markov times τ_t, $n = 1,2,\cdots$ (relative to $F = \{\mathscr{F}_t\}$, $t \in \boldsymbol{T}$), we have $x_{\tau_n(\omega)}(\omega) \to x_{\tau(\omega)}(\omega)$ $\mathbf{P}_x$-a.s. on the set $\{\tau < \infty\}$ for all $x \in E$ as $n \to \infty$, where $\tau = \lim_{n\to\infty} \tau_n$.

We note that the requirement of progressive measurability involved in these definitions guarantees $\mathscr{F}_\tau/\mathscr{B}$-measurability of the variables $x_\tau(\omega)$ and $x_{\tau_n}(\omega)$ (see Lemma 9, § 2).

2. *Transition function.* Let us write $P(t,x,\Gamma)$ for $\mathbf{P}_x\big(x_t(\omega) \in \Gamma\big)$, $x \in E$, $\Gamma \in \mathscr{B}$, $t \in \boldsymbol{Z}$. The function $P(t,x,\Gamma)$ is called the *transition function* of the Markov process X. Its properties below follow directly from Definition 1:

1) $P(t,x,\cdot)$ is a measure on $E(\ ,\mathscr{B})$ for all $x \in E$ and $t \in \boldsymbol{Z}$.

2) $P(t,x,\Gamma)$ is a $\mathscr{B}$-measurable function of x for all $t \in \boldsymbol{Z}$ and $\Gamma \in \mathscr{B}$.

3) (Kolmogorov-Chapman equation)

$$P(t+s,x,\Gamma) = \int_E P(s,x,dy)\, P(t,y,\Gamma) \qquad t, s \in \boldsymbol{Z}. \tag{1.11}$$

4) $P(0,x,\Gamma) = \chi_\Gamma(x)$.

For the case of discrete time $t \in \boldsymbol{Z}$, in view of (1.11) the transition function $P(t,x,\Gamma)$ is completely determined by the one step transition function $P(x,\Gamma) = P(1,x,\Gamma)$.

3. *Standard processes.* In the case of continuous time the definition of Markov process given above turns out to be too broad to construct a productive theory. In this section we examine the important concept of a standard Markov process; in Chapter III we investigate optimal stopping problems for such a process.

We assume that the original spaces $(E, \mathscr{B})$ and $(\Omega, \mathscr{F})$ have the following structure:

E is a locally compact separable metric space with metric $d(\cdot,\cdot)$; (1.12)

$\mathscr{B}$ is the σ-algebra of subsets of E generated by open sets; (1.13)

Ω is the set of functions $\omega = \omega(t)$, $t \in \boldsymbol{T}$, with values in E, right continuous and having limits from the left; (1.14)

$\mathscr{F}$ is the σ-algebra of ω-sets generated by sets of the form

$$\{\omega\colon \omega(s) \in \Gamma\}, \qquad \Gamma \in \mathscr{B},\ s \in \boldsymbol{T}. \tag{1.15}$$

For each $\omega \in \Omega$ we let $x_t(\omega) = \omega(t)$ and call $\{x_t(\omega)\}$, $t \in \boldsymbol{T}$, the trajectory corresponding to the elementary event ω; moreover, let $\mathscr{F}_t = \sigma\{\omega\colon x_s(\omega),\ s \leqq t\}$.

Using the assumptions (1.12)—(1.15), it can easily be shown that for each t the σ-algebra $\mathscr{F}_t$ can be generated by a countable system of sets of the form $\{\omega\colon x_r(\omega) \in \Gamma_i\}$, where r are rational numbers in $[0,t]$ and Γ_i is an element of the base $\{\Gamma_1,\Gamma_2,\cdots\}$, consisting of open sets of the space E.

In many problems in the theory of Markov processes, the σ-algebras $\mathscr{B}$, $\mathscr{F}$ and $\mathscr{F}_t$, $t \in \boldsymbol{T}$, turn out to be too restricted (see, for example, Theorem 2 in § 2), and we need to introduce their completions $\bar{\mathscr{B}}$, $\bar{\mathscr{F}}$ and $\bar{\mathscr{F}}_t$, obtained as follows.

Let μ be a probability measure on $(E,\mathscr{B})$, let $\mathscr{B}^\mu$ be the completion of the σ-algebra $\mathscr{B}$ with respect to the measure μ, and let $\mathbf{P}^\mu(A) = \int_E \mathbf{P}_x(A)\mu(dx)$, $A \in \mathscr{F}$. We let (see § 1) $\bar{\mathscr{B}} = \bigcap_\mu \mathscr{B}^\mu$, $\bar{\mathscr{F}} = \bigcap_\mu \mathscr{F}^{\mathbf{P}^\mu}$ and $\bar{\mathscr{F}}_t = \bigcap_\mu \mathscr{F}_t^{\mathbf{P}^\mu}$, and we let $\bar{\mathbf{P}}_x$ denote the extension of the measure $\mathbf{P}_x$ to $\bar{\mathscr{F}}$.

It is known ([**25**], Theorem 3.12) that if the process $X = (x_t,\mathscr{F}_t,\mathbf{P}_x)$ is strong Markov and quasi-left-continuous, and $\mathscr{F}_t = \mathscr{F}_{t+0}$, then the process $X = (x_t,\bar{\mathscr{F}}_t, \bar{\mathbf{P}}_x)$ is also strong Markov and quasi-left-continuous in the phase space $(E, \bar{\mathscr{B}})$. Therefore, considering strong Markov quasi-left-continuous Markov processes satisfying the condition $\mathscr{F}_t = \mathscr{F}_{t+0}$, we can immediately assume that

$$\mathscr{B} = \bar{\mathscr{B}}, \quad \mathscr{F} = \bar{\mathscr{F}}, \quad \mathscr{F}_t = \bar{\mathscr{F}}_t, \quad \bar{\mathbf{P}}_x = \mathbf{P}_x. \tag{1.16}$$

DEFINITION 4. A strong Markov, quasi-left-continuous Markov process $X = (x_t,\mathscr{F}_t,\mathbf{P}_x)$, $t \in \boldsymbol{T}$, is said to be *standard* if $\mathscr{F}_t = \mathscr{F}_{t+0}$ for all t and conditions (1.12)—(1.15), (1.16) are satisfied.

4. *The semigroup* $\{T_t\}$. *Feller processes.* Let $B(E,\mathscr{B})$ be the space of $\mathscr{B}$-measurable bounded functions $f(x)$, $x \in E$, with norm $\|f\| = \sup_{x \in E} |f(x)|$. With each function $f \in B(E,\mathscr{B})$ we associate a function

$$T_t f(x) = \int_E f(y)P(t, x, dy), \qquad t \in \boldsymbol{Z}. \tag{1.17}$$

Formula (1.17) defines a family of linear operators $\{T_t\}$, $t \in \boldsymbol{Z}$. By (1.11), this family forms a semigroup, i.e.

$$T_s \cdot T_t = T_{s+t}, \qquad s, t \geqq 0.$$

It is easy to see that this semigroup is contracting: $\|T_t f\| \leqq \|f\|$, $t \geqq 0$.

Let $C(E,\mathscr{B}) \subseteq B(E,\mathscr{B})$ be the space of bounded $\mathscr{B}$-measurable continuous functions given on the space $(E,\mathscr{B})$ which satisfies conditions (1.12) and (1.13).

DEFINITION 5. The semigroup of operators $\{T_t\}$, $t \in \boldsymbol{T}$, is called a *Feller semigroup* (also the corresponding transition function $P(t,x,\Gamma)$ is called a *Feller transition function* and the corresponding Markov process X a *Feller process*) if for each $f \in C(E,\mathscr{B})$ the function $T_t f(x)$ is continuous in $x \in E$ for $t \in \boldsymbol{T}$.

It is known ([**25**], Theorem 3.3) that if the Markov process $X = (x_t,\mathscr{F}_t,\mathbf{P}_x)$, $t \in \boldsymbol{T}$, is a Feller process and conditions (1.12) and (1.13) are satisfied, then the process $X' = (x_t,\mathscr{F}_{t+0},\mathbf{P}_x)$, $t \in \boldsymbol{T}$, is also Markov. Indeed, if we consider Feller processes whose phase space satisfies conditions (1.12) and (1.13), then without loss of generality we can assume satisfaction of the condition

$$\mathscr{F}_t = \mathscr{F}_{t+0}, \qquad t \in \boldsymbol{T}, \tag{1.18}$$

involved in the definition of a standard process.

5. *The operators θ_t and θ_τ.* Let $X = (x_t,\mathscr{F}_t,\mathbf{P}_x)$, $t \in \boldsymbol{Z}$, be a Markov process for which $\Omega = E^{\boldsymbol{Z}}$.

DEFINITION 6. Suppose that for all $t \in \boldsymbol{Z}$ and $\omega \in \Omega$, $\theta_t\omega$ is an element of the space Ω such that

$$x_s(\theta_t\omega) = x_{s+t}(\omega) \qquad \text{for all } s \in \boldsymbol{Z}. \tag{1.19}$$

If $f = f(\omega)$ is a function on Ω, then $\theta_t f = \theta_t f(\omega)$ denotes the function $f(\theta_t\omega)$.

If $\tau = \tau(\omega)$ is an $\mathscr{F}$-measurable function with values in $\boldsymbol{Z}$, then $\theta_{\tau(\omega)}\omega$ denotes the element of Ω such that $\theta_{\tau(\omega)}\omega = \theta_t\omega$ if $\tau(\omega) = t$. By $\theta_\tau f$ we mean the function $f(\theta_\tau\omega)$.

With the aid of the new operator θ_τ the strong Markov property (1.10) can be written in the following equivalent form:[9] if $\eta = \eta(\omega)$ is an $\mathscr{F}$-measurable function such that $\mathbf{E}_x|\eta(\omega)| < \infty$, and $\tau = \tau(\omega)$ is a Markov time (relative to the system $\{\mathscr{F}_t\}$, $t \in \boldsymbol{Z}$, $\mathscr{F}_t = \sigma(\omega\colon \omega(s), s \leqq t)$), then on the set $\{\omega\colon \tau(\omega) < \infty\}$

$$\mathbf{E}_x\{\theta_\tau\eta \mid \mathscr{F}_\tau\} = \mathbf{E}_{x_\tau}\eta \qquad (\mathbf{P}_x\text{-a.s., } x \in E). \tag{1.20}$$

From (1.20) it follows that if the random variable $\xi = \xi(\omega)$ is $\mathscr{F}_\tau$-measurable and $\mathbf{E}_x|\xi_\tau| < \infty$ and $\mathbf{E}_x|\xi\theta_\tau\eta| < \infty$, $x \in E$, then

$$\mathbf{E}_x\{\xi\theta_\tau\eta\} = \mathbf{E}_x\{\xi\mathbf{E}_{x_\tau}\eta\}. \tag{1.21}$$

6. *Infinitesimal and characteristic operators.* Let $(E,\mathscr{B})$ be a phase space and

[9] We assume that $\Omega = E^{\boldsymbol{Z}}$ and $\mathscr{F} = \sigma\{\omega : \omega(s), s \in \boldsymbol{Z}\}$.

$B(E,\mathscr{B})$ be the Banach space of bounded measurable functions $f = f(x)$ with norm $\| f \| = \sup_{x \in E} |f(x)|$.

The *infinitesimal operator* $\mathscr{A}$ *of the semigroup* $\{T_t\}, t \in \boldsymbol{T}$, is defined by the formula

$$\mathscr{A}f(x) = \lim_{t \to 0} \frac{T_t f(x) - f(x)}{t}. \tag{1.22}$$

In order to completely determine the operator $\mathscr{A}$, we have to give its domain of definition $\mathscr{D}_{\mathscr{A}}$. We assume that $\mathscr{D}_{\mathscr{A}}$ consists of all functions $f \in B(E,\mathscr{B})$ for which the limit on the right side of (1.22) is uniform over $x \in E$.

If $\{T_t\}$, $t \in \boldsymbol{T}$, is a semigroup corresponding to a Markov process $X = \{x_t, \mathscr{F}_t, \mathbf{P}_x\}$, then $\mathscr{A}$ is said to be the *infinitesimal operator of the process* X.

The *weak infinitesimal operator* $\tilde{\mathscr{A}}$ is defined, like the operator $\mathscr{A}$, by formula (1.22) but with the wider domain of definition $\mathscr{D}_{\tilde{\mathscr{A}}}$. With $\mathscr{D}_{\tilde{\mathscr{A}}}$ are associated those functions from $B(E,\mathscr{B})$ for which: a) the ratio an the right side of (1.22) is bounded for all $x \in E$ and t from some neighborhood of zero; b) the limit of this ratio exists for each $x \in E$ and determines a function $(\tilde{\mathscr{A}}f(x))$ for which $T_t \tilde{\mathscr{A}}f(x)$ converges weakly to $\tilde{\mathscr{A}}f(x)$ as $t \to 0$ (see [25], Chapter II, § 2.9). The domain of definition $\mathscr{D}_{\tilde{\mathscr{A}}} \supseteq \mathscr{D}_{\mathscr{A}}$.

We let $\mathscr{U}$ denote the collection of all open sets (in the topology generated by the metric $d(\cdot,\cdot)$) having compact closures.[10]

Suppose $X = (x_t, \mathscr{F}_t, \mathbf{P}_x)$, $t \in \boldsymbol{T}$, $x \in E$, is a standard Markov process and

$$\sigma(U) = \inf \{t \geqq 0: x_t(\omega) \in E \backslash U\}, \qquad U \in \mathscr{U}.$$

By Lemma 4.1 of [25] (also compare Theorem 2 of § 2), the time $\sigma(U)$ is Markov.

Let $f(x)$ be an arbitrary $\bar{\mathscr{B}}$-measurable funciion and let $\tilde{\mathscr{U}}$ be the collection of sets $U \in \mathscr{U}$ for which[11] $\mathbf{E}_{x_0} |f(x_{\sigma(U)})| < \infty$, $x_0 \in E$.

We form the expression

$$\left(\mathbf{E}_{x_0} f(x_{\sigma(U)}) - f(x_0)\right) / \mathbf{E}_{x_0} \sigma(U), \tag{1.23}$$

and assume it equal to zero if $\mathbf{E}_{x_0} \sigma(U) = \infty$, and set

$$\mathfrak{A} f(x_0) = \lim_{U \downarrow x_0} \frac{\mathbf{E}_{x_0} f(x_{\sigma(U)}) - f(x_0)}{\mathbf{E}_{x_0} \sigma(U)}, \tag{1.24}$$

where the limit is taken over the system of neighborhoods $U \in \tilde{\mathscr{U}}$ shrinking to the point x_0 (for more detail see [25], Chapter V, § 3).

The set of all $\bar{\mathscr{B}}$-measurable functions for which the limit (1.24) exists at the point x_0 is denoted by $\mathscr{D}_{\mathfrak{A}}(x_0)$. If $f \in \mathscr{D}_{\mathfrak{A}}(x_0)$ for all $x_0 \in G$, then we write $f \in \mathscr{D}_{\mathfrak{A}}(G)$. In the case when $G = E$ we let $\mathscr{D}_{\mathfrak{A}} = \mathscr{D}_{\mathfrak{A}}(E)$.

For a broad class of Markov processes the characteristic operator is an extension

---

[10] The closure of a set $U \in \mathscr{U}$ is the smallest closed set containing U.

[11] By $\mathbf{E}_{x_0} f(x_{\sigma(U)})$ we mean the integral $\int_{(\sigma(U) < \infty)} f(x_{\sigma(U)}) d\mathbf{P}_{x_0}$.

of the weak infinitesimal operator: $\mathscr{D}_{\mathfrak{A}} \supseteq \mathscr{D}_{\tilde{\mathscr{A}}}$ (see for example, Theorem 5.5. in **[25]**).

7. *Terminating inhomogeneous Markov processes.* Optimal stopping problems will be further investigated primarily for homogeneous nonterminating Markov processes. However, the theory presented can also be applied with hardly any change in the case of both terminating and inhomogeneous Markov processes. We give the definitions needed for these processes.

Let $(E,\mathscr{B})$ be a phase space. Let $E_\Delta = E \cup \{\Delta\}$, where Δ is a ("fictitious") point not belonging to E, and suppose $\mathscr{B}_\Delta$ is the σ-algebra of subsets of E_Δ generated by sets from $\mathscr{B}$. We note that $\{\Delta\} \in \mathscr{B}_\Delta$, so that the space $(E_\Delta,\mathscr{B}_\Delta)$ is a phase space.

We use $\Omega = E_\Delta^{\bar{\boldsymbol{Z}}}$, $\bar{\boldsymbol{Z}} = \boldsymbol{Z} \cup \{\infty\}$, to denote the space of functions $\omega = \omega(t)$, $t \in \bar{\boldsymbol{Z}}$, with values in E_Δ such that $\omega(\infty) = \Delta$ and $\omega(t) = \Delta$ for all $t \geqq s$ if $\omega(s) = \Delta$.

Let $\omega_\Delta = \omega_\Delta(t)$ be a function such that

$$\omega_\Delta(t) \equiv \Delta, \quad t \in \bar{\boldsymbol{Z}}, \quad \mathscr{F}_t^s = \sigma\{\omega\colon \omega(u),\ s \leqq u \leqq t\},$$
$$\mathscr{F}_t = \mathscr{F}_t^0, \quad \mathscr{F}^s = \mathscr{F}_\infty^s, \quad \mathscr{F} = \mathscr{F}_\infty^0, \quad x_t(\omega) = \omega(t), \quad t \in \bar{\boldsymbol{Z}},$$
$$\zeta(\omega) = \inf\{t \geqq 0\colon x_t(\omega) = \Delta\}.$$

The variable $\zeta(\omega)$ is called the *life time* or *time of termination* (of the trajectory $\omega = \omega(t)$, $t \in \bar{\boldsymbol{Z}}$). Since

$$\{\omega\colon \zeta(\omega) < t\} = \bigcup_{r<t} \{\omega(r) = \Delta\} \in \mathscr{F}_t$$

(r are rational numbers), $\zeta(\omega)$ is an $\mathscr{F}$-measurable function, i.e. it is a real random variable with values in $\boldsymbol{Z}$.

Now suppose that for all $x \in E_\Delta$ and $t \in \boldsymbol{Z}$ we are given probability measures $\mathbf{P}_{s,x}$ on sets from $\mathscr{F}^s$.

DEFINITION 7. The system $X = (x_t,\zeta(\omega),\mathscr{F}_t^s,\mathbf{P}_{s,x})$, s, $t \in \boldsymbol{Z}$, is called a (inhomogeneous, terminating) *Markov process in the phase space* $(E,\mathscr{B})$ *with adjoined point* $\{\Delta\}$ if the following conditions are satisfied:

1) $\mathbf{P}_{s,x}(A)$ is a $\mathscr{B}$-measurable function of x for each $A \in \mathscr{F}$ and $s \in \boldsymbol{Z}$.

2) For all $x \in E_\Delta$, $B \in \mathscr{B}_\Delta$, $0 \leqq s \leqq t \leqq u$,

$$\mathbf{P}_{s,x}(x_u \in B \mid \mathscr{F}_t^s) = P_{t,x_t}(x_u \in B) \qquad (\mathbf{P}_{s,x}\text{-a. s.}). \tag{1.25}$$

3) $\mathbf{P}_{s,x}(x_s = x) = 1, \quad x \in E_\Delta$.

The function $P(s; x; t,\Gamma) = \mathbf{P}_{s,x}(x_t(\omega) \in \Gamma)$ is called the *transition function* of the Markov process.

Analogously to Definition 2, we introduce the concept of a strong Markov inhomogeneous terminating process (for more detail see **[24]**, Chapter 5). We restrict consideration to homogeneous terminating processes.

The strong Markov property in this case is stated as in (1.10). The only change is that instead of the set $\{\omega: \tau(\omega) < \infty\}$ we have to use the set $\{\omega: \tau(\omega) < \zeta(\omega)\}$. Analogous changes also have to be made in the definition of a quasi-continuous process.

In the definition of a standard process, for the space Ω we have to take the set of functions $\omega = \omega(t)$, $t \in \bar{T}$, with values in E_Δ, right continuous, having limit from the left for $t < \zeta(\omega)$, and such that if $\omega(t) = \Delta$, then $\omega(u) = \Delta$ for $u \geqq t$. In the definition of a standard process we also include the assumption

$$\lim_{t \to 0} \mathbf{P}_x(x_t \in E) = 1, \qquad x \in E.$$

The concepts of semigroup, infinitesimal and characteristic, and also the operators θ_t and θ_τ, also carry over to the case of terminating homogeneous processes ([**24**]).

§ 4. Martingales and super-martingales

1. The structure of the theory of optimal stopping rules (for both Markov and other processes) is essentially based on properties of martingales and semimartingales. We spend some time on the basic definitions and some results obtained in subsequent chapters.

Let $(\Omega, \mathscr{F}, \mathbf{P})$ be a probability space. Suppose that for each[12] $t \in \boldsymbol{Z}$ we are given the σ-algebras $\mathscr{F}_t \subseteq \mathscr{F}$ and $\mathscr{F}_t$-measurable random variables $x_t = x_t(\omega)$ having the following properties:

$$\mathscr{F}_s \subseteq \mathscr{F}_t \quad \text{whenever } s \leqq t, \tag{1.26}$$

$$\mathbf{E}\,|\,x_t\,| < \infty \quad \text{for all } t \in \boldsymbol{Z}, \tag{1.27}$$

$$\mathbf{E}(x_t | \mathscr{F}_s) \leqq x_s \quad \text{whenever } s \leqq t \quad (\mathbf{P}\text{-a.s.}). \tag{1.28}$$

DEFINITION 1. A system $X = (x_t, \mathscr{F}_t, \mathbf{P})$, $t \in \boldsymbol{Z}$, satisfying conditions (1.26)—(1.28) is called a *super-martingale*. If instead of (1.28) the condition

$$\mathbf{E}(x_t \,|\, \mathscr{F}_s) \geqq x_s \quad \text{whenever } s \leqq t \quad (\mathbf{P}\text{-a.s.}) \tag{1.29}$$

is satisfied, then X is called a *sub-martingale*; but if the condition

$$\mathbf{E}(x_t \,|\, \mathscr{F}_s) = x_s \quad \text{whenever } s \leqq t \quad (\mathbf{P}\text{-a.s.}) \tag{1.30}$$

is satisfied, then it is a *martingale*.

Obviously if X is a super-martingale then the process $-X$ is a sub-martingale. Therefore in studying their properties it suffices to consider, say, super-martingales.

DEFINITION 2. If instead of (1.27) one of the conditions

---

[12] Recall that $\boldsymbol{Z} = \boldsymbol{T} = [0, \infty)$ in the case of continuous time and $\boldsymbol{Z} = \boldsymbol{N} = \{0, 1, \cdots\}$ in the case of discrete time.

$$\mathbf{E}x_t^+ < \infty \text{ or } \mathbf{E}x_t^- < \infty \tag{1.31}$$

(guaranteeing the existence of $\mathbf{E}x_t$) is satisfied, then the system X satisfying conditions (1.26) and (1.28), (1.29) or (1.30) is called a *generalized super-martingale, sub-martingale* or *martingale,* respectively.

2. In the case of continuous time $t \in \boldsymbol{T}$, in formulating the properties of martingales we assume (without especially saying it every time) that with probability one their trajectories are right continuous and $\mathscr{F} = \mathscr{F}^{\mathbf{P}}$ and $\mathscr{F}_t = \mathscr{F}_t^{\mathbf{P}}$, $t \in \boldsymbol{T}$.

THEOREM 3. *Suppose that $\bar{\boldsymbol{Z}} = \boldsymbol{Z} \cup \{\infty\}$ and the system $X = (x_t, \mathscr{F}_t, \mathbf{P})$, $t \in \boldsymbol{Z}$, such that $\sup_{t \in \boldsymbol{Z}} \mathbf{E}x_t^-(\omega) < \infty$, forms a super-martingale (martingale). Then with probability one the limit $x_\infty(\omega) = \lim_{t\to\infty} x_t(\omega)$ exists and is finite.*

If the random variables $\{x_t\}$, $t \in \boldsymbol{Z}$, are uniformly integrable,[13] *then the system $X = (x_t, \mathscr{F}_t, \mathbf{P})$, $t \in \bar{\boldsymbol{Z}}$, where $\mathscr{F}_\infty = \sigma(\bigcup_{t \in \boldsymbol{Z}} \mathscr{F}_t)$, $x_\infty = \lim_{t\to\infty} x_t$, forms a super-martingale (martingale).*

Let $X = (x_t, \mathscr{F}_t, \mathbf{P})$, $t \in \boldsymbol{Z}$, be a generalized super-martingale. Then the limit $x_\infty(\omega) = \lim_{t\to\infty} x_t(\omega)$ exists, and is finite or equal to $+\infty$ for almost all ω such that $\inf_{s \in \boldsymbol{Z}} \sup_{t \in \boldsymbol{Z}} \mathbf{E}(x_t^- \mid \mathscr{F}_s) < \infty$.

From this theorem we can derive the following important result on properties of conditional mathematical expectations.

THEOREM 4. *Suppose that in the probability space $(\Omega, \mathscr{F}, \mathbf{P})$ we are given a nondecreasing sequence of σ-algebras $\mathscr{F}_t \subseteq \mathscr{F}$, $t \in \boldsymbol{Z}$, $\mathscr{F}_s \subseteq \mathscr{F}_t$, $s \leqq t$, and $\eta = \eta(\omega)$ is a random variable such that $\mathbf{E}\,|\,\eta(\omega)\,| < \infty$. Then with probability one*

$$\lim_{t\to\infty} \mathbf{E}(\eta(\omega) \mid \mathscr{F}_t) = \mathbf{E}(\eta(\omega) \mid \mathscr{F}_\infty) \qquad (\mathbf{P}\text{-a.s.}), \tag{1.32}$$

where $\mathscr{F}_\infty = \sigma(\bigcup_{t \in \boldsymbol{Z}} \mathscr{F}_t)$.

If the random variable $\eta = \eta(\omega)$ is such that $\mathbf{E}\eta^+(\omega) < \infty$, then

$$\lim_{t\to\infty} \mathbf{E}(\eta(\omega) \mid \mathscr{F}_t)$$

exists with probability one, and

$$\lim_{t\to\infty} \mathbf{E}(\eta(\omega) \mid \mathscr{F}_t) \leqq \mathbf{E}(\eta(\omega) \mid \mathscr{F}_\infty) \quad (\mathbf{P}\text{-a.s.}).$$

THEOREM 5. *Let $X = (x_t, \mathscr{F}_t, \mathbf{P})$, $t \in \boldsymbol{Z}$, be a super-martingale, and suppose there exists an $\mathscr{F}$-measurable random variable $\eta = \eta(\omega)$ such that*

$$\mathbf{E}\,|\,\eta(\omega)\,| < \infty,\ x_t(\omega) \geqq \mathbf{E}(\eta \mid \mathscr{F}_t), \qquad t \in \boldsymbol{Z} \qquad (\mathbf{P}\text{-a.s.}). \tag{1.33}$$

If $\tau = \tau(\omega)$ and $\sigma = \sigma(\omega)$ are Markov times (relative to $F = \{\mathscr{F}_t\}$, $t \in \boldsymbol{Z}$), $\tau \geqq \sigma$ ($\mathbf{P}$-a.s.) and $\mathbf{P}(\tau < \infty) = 1$, then the random variables x_σ, x_τ are integrable and

[13] A sequence of random variables $\{x_t\}$, $t \in \mathbf{Z}$, is said to be *uniformly integrable* if $\lim_{K\to\infty} \mathbf{E}\{|x_t|;\, |x_t| < K\} = 0$ uniformly in $t \in \mathbf{Z}$. For such sequences $\sup_{t \in \mathbf{Z}} \mathbf{E}|x|_t < \infty$.

$$x_\sigma \geqq \mathbf{E}(x_\tau \mid \mathscr{F}_\sigma) \qquad (\mathbf{P}\text{-}a.s.). \tag{1.34}$$

If, in particular, $X = (x_t,\mathscr{F}_t,\mathbf{P})$, $t \in \boldsymbol{Z}$, *is a uniformly integrable martingale, then*

$$x_\sigma = \mathbf{E}(x_\tau \mid \mathscr{F}_\sigma) \qquad (\mathbf{P}\text{-}a.s.). \tag{1.35}$$

The inequality (1.34) plays a central role in the construction of the theory of optimal stopping rules. Therefore we go into its proof, restricting consideration to the case of discrete time $\boldsymbol{Z} = \boldsymbol{N}$.

First we suppose that $\mathbf{P}(\tau \leqq N) = 1$ for some $N < \infty$. The function x_τ is obviously $\mathscr{F}_\tau$-measurable, $\mathbf{E}\,|\,x_\tau\,| < \infty$, and consequently the conditional mathematical expectation $\mathbf{E}(x_\tau \mid \mathscr{F}_\sigma)$ is defined.

Suppose the difference $\tau - \sigma$ takes on at most one of the two values 0 and 1. Then for each $\Lambda \in \mathscr{F}_\sigma$

$$\int_\Lambda (x_\sigma - x_\tau)\, d\mathbf{P} = \sum_{n=0}^{N} \int_{\Lambda \cap \{\sigma=n\} \cap \{\tau>n\}} (x_n - x_{n+1})\, d\mathbf{P}.$$

The events $\Lambda \cap \{\sigma = n\} \in \mathscr{F}_n$ and $\{\tau > n\} = \overline{\{\tau \leqq n\}} \in \mathscr{F}_n$. Therefore, by (1.28), for every $n = 0,1,\cdots,N$

$$\int_{\Lambda \cap \{\sigma=n\} \cap \{\tau>n\}} (x_n - x_{n+1})\, d\mathbf{P} \geqq 0$$

and consequently

$$\int_\Lambda x_\sigma\, d\mathbf{P} \geqq \int_\Lambda x_\tau\, d\mathbf{P}.$$

Hence obviously $x_\sigma \geqq \mathbf{E}(x_\tau \mid \mathscr{F}_\sigma)$ ($\mathbf{P}$-a.s.).

In order to get rid of the assumption that the difference takes on only the two values 0 and 1, we let

$$\tau_n = \min(\tau,\ \sigma + n), \qquad n = 0,\ 1,\cdots,N.$$

For each n the time τ_n is Markov, and the difference $\tau_{n+1} - \tau_n$ takes on only the values 0 and 1.

Suppose $\Lambda \in \mathscr{F}_\sigma$. Then $\Lambda \in \mathscr{F}_{\tau_n}$ for all $n = 0,1,\cdots,N$ and, by the reasoning above,

$$\int_\Lambda x_\sigma d\mathbf{P} \geqq \int_\Lambda x_{\tau_1} d\mathbf{P} \geqq \cdots \geqq \int_\Lambda x_{\tau_N}\, d\mathbf{P} = \int_\Lambda x_\tau d\mathbf{P},$$

which also proves (1.34) for the case where $\mathbf{P}(\tau \leqq N) = 1$, $N < \infty$.

For the proof of (1.34) in the general case we represent x_n in the following form:

$$x_n = \mathbf{E}(\eta \mid \mathscr{F}_n) + \big(x_n - \mathbf{E}(\eta \mid \mathscr{F}_n)\big).$$

Let $\zeta_n = \mathbf{E}(\eta \mid \mathscr{F}_n)$ and $\gamma_n = x_n - \mathbf{E}(\eta \mid \mathscr{F}_n)$. Clearly $(\zeta_n,\mathscr{F}_n,\mathbf{P})$, $n \in \boldsymbol{N}$, forms a martingale, and $(\gamma_n, \mathscr{F}_n, \mathbf{P})$, $n \in \boldsymbol{N}$, is a nonnegative super-martingale.

We establish the relationships

$$\zeta_\sigma = \mathbf{E}(\zeta_\tau \mid \mathscr{F}_\sigma), \qquad \gamma_\sigma \geqq \mathbf{E}(\gamma_\tau \mid \mathscr{F}_\sigma) \qquad (\mathbf{P}\text{-a.s.}),$$

from which the desired inequality (1.34) follows in an obvious fashion.

For proof of the first equality (cf. (1.35)) it is sufficient to establish that for each Markov time τ^* such that $\mathbf{P}(\tau^* < \infty) = 1$ the equality

$$\zeta_{\tau^*} = \mathbf{E}(\eta \mid \mathscr{F}_{\tau^*})$$

is satisfied. Indeed, then we immediately obtain

$$\zeta_\sigma = \mathbf{E}(\eta \mid \mathscr{F}_\sigma) = \mathbf{E}[\mathbf{E}(\eta \mid \mathscr{F}_\tau) \mid \mathscr{F}_\sigma] = \mathbf{E}(\zeta_\tau \mid \mathscr{F}_\sigma).$$

Let $\Lambda \in \mathscr{F}_{\tau^*}$ and $\tau_k^* = \min(\tau^*, k)$, $k \in \boldsymbol{N}$. The set $\Lambda \cap \{\tau^* \leqq k\} \in \mathscr{F}_{\tau_k^*}$ and, as we established above,

$$\int_{\Lambda \cap \{\tau^* \leqq k\}} \zeta_{\tau_k^*} d\mathbf{P} = \int_{\Lambda \cap \{\tau^* \leqq k\}} \zeta_k d\mathbf{P} = \int_{\Lambda \cap \{\tau^* \leqq k\}} \eta d\mathbf{P}.$$

Since the sequence of random variables $\{\zeta_{\tau_k^*}\}$, $k \in \boldsymbol{N}$, is uniformly integrable (see, for example, [49], Chapter V, Theorem 19), on passing to the limit $(k \to \infty)$ in the preceding equality we obtain

$$\int_\Lambda \zeta_{\tau^*} d\mathbf{P} = \int_\Lambda \eta d\mathbf{P},$$

which proves the desired relationship $\zeta_{\tau^*} = \mathbf{E}(\eta \mid \mathscr{F}_{\tau^*})$ ($\mathbf{P}$-a.s.).

For the proof of the inequality $\gamma_\sigma \geqq \mathbf{E}(\gamma_\tau \mid \mathscr{F}_\sigma)$ ($\mathbf{P}$-a.s.) we set

$$\sigma_k = \min(\sigma, k), \qquad \tau_k = \min(\tau, k), \qquad k \in \boldsymbol{N}.$$

It is easy to see that the sequences $\{\gamma_{\tau_k}, \mathscr{F}_{\tau_k}, \mathbf{P}\}$ and $\{\gamma_{\sigma_k}, \mathscr{F}_{\sigma_k}, \mathbf{P}\}$, $k \in \boldsymbol{N}$, are nonnegative super-martingales. Therefore, by Theorem 3,

$$\lim_{k\to\infty} \gamma_{\tau_k} = \gamma_\tau \quad \text{and} \quad \lim_{k\to\infty} \gamma_{\sigma_k} = \gamma_\sigma \ (\mathbf{P}\text{-a.s.}).$$

Since $\mathbf{E}\gamma_{\tau_k} \leqq \mathbf{E}\gamma_0 < \infty$, by Fatou's lemma

$$\mathbf{E}\gamma_\tau \leqq \varliminf_{k\to\infty} \mathbf{E}\gamma_{\tau_k} \leqq \mathbf{E}\gamma_0 < \infty.$$

Hence it follows that the nonnegative variable γ_τ (and analogousy γ_σ) is integrable.

Let $\Lambda \in \mathscr{F}_\sigma$. Then $\Lambda \cap \{\sigma \leqq k\} \in \mathscr{F}_{\sigma_k}$ and, as shown above,

$$\int_{\Lambda \cap \{\sigma \leqq k\}} \gamma_{\sigma_k} d\mathbf{P} \geqq \int_{\Lambda \cap \{\sigma \leqq k\}} \gamma_{\tau_k} d\mathbf{P}.$$

But since $\{\sigma \leqq k\} \supseteq \{\tau \leqq k\}$, we have

$$\int_{\Lambda \cap \{\sigma \leqq k\}} \gamma_{\sigma_k} d\mathbf{P} \geqq \int_{\Lambda \cap \{\tau \leqq k\}} \gamma_{\tau_k} d\mathbf{P}$$

or

$$\int_{\Lambda \cap \{\sigma \leqq k\}} \gamma_{\sigma} d\mathbf{P} \geqq \int_{\Lambda \cap \{\tau \leqq k\}} \gamma_{\tau} d\mathbf{P}.$$

Letting $k \to \infty$, we now obtain the inequality

$$\int_{\Lambda} \gamma_{\sigma} d\mathbf{P} \geqq \int_{\Lambda} \gamma_{\tau} d\mathbf{P}, \qquad \Lambda \in \mathscr{F}_{\sigma},$$

from which it follows that $\gamma_{\sigma} \geqq \mathbf{E}(\gamma_{\tau} \mid \mathscr{F}_{\sigma})$ ($\mathbf{P}$-a.s.).

We note that the assertion of the theorem also remains valid without the assumption $\mathbf{P}(\tau < \infty) = 1$ (cf. Lemma 1 in Chapter II).[14] In this case, x_{∞} in (1.34) means the limit $\lim_{t \to \infty} x_t$, which, according to Theorem 3, exists, since $\sup_{t \in \boldsymbol{Z}} \mathbf{E} x_t^- < \infty$ by (1.33).

It follows from (1.35) that if $X = (x_t, \mathscr{F}_t, \mathbf{P})$, $t \in \boldsymbol{Z}$, is a uniformly integrable martingale, then for any Markov time τ such that $\mathbf{P}(\tau < \infty) = 1$,

$$\mathbf{E} x_{\tau} = \mathbf{E} x_0. \tag{1.36}$$

In the next theorem we give conditions for satisfaction of the equality (1.36) for a given Markov time $\tau = \tau(\omega)$ satisfying the condition $\mathbf{P}(\tau < \infty) = 1$ without the assumption of uniform integrability of the martingale X.

THEOREM 6. *Let $X = (x_t, \mathscr{F}_t, \mathbf{P})$, $t \in \boldsymbol{Z}$, be a martingale, and let $\tau = \tau(\omega)$ be a Markov time such that $\mathbf{P}(\tau < \infty) = 1$. For satisfaction of the equality* (1.36) *it is sufficient to have satisfaction of the following two conditions*:

$$\mathbf{E} |x_{\tau}| < \infty, \tag{1.37}$$

$$\lim_{t \to \infty} \int_{\{\tau > t\}} x_t d\mathbf{P} = 0. \tag{1.38}$$

The proof of this theorem for the case of discrete time $t \in \boldsymbol{N}$ is extremely simple. Indeed, since $\mathbf{E} |x_{\tau}| < \infty$, for all $n \in \boldsymbol{N}$ we have

$$\begin{aligned} \mathbf{E} x_{\tau} &= \int_{\Omega} x_{\tau} d\mathbf{P} = \int_{\{\tau \leqq n\}} x_{\tau} d\mathbf{P} + \int_{\{\tau > n\}} x_{\tau} d\mathbf{P} \\ &= \sum_{k=0}^{n} \int_{\{\tau = k\}} x_k d\mathbf{P} + \int_{\{\tau > n\}} x_{\tau} d\mathbf{P} = \sum_{k=0}^{n} \int_{\{\tau = k\}} x_n d\mathbf{P} + \int_{\{\tau > n\}} x_{\tau} d\mathbf{P} \\ &= \int_{\{\tau \leqq n\}} x_n d\mathbf{P} + \int_{\{\tau > n\}} x_{\tau} d\mathbf{P} \end{aligned}$$

---

[14] The lemmas and theorems are numbered independently in each chapter of the book. In references to theorems and lemmas from other chapters we use a double numbering (for example, Theorem II. 1 indicates the first theorem of the second chapter).

$$= \int_{\Omega} x_n \, d\mathbf{P} - \int_{\{\tau > n\}} x_n \, d\mathbf{P} + \int_{\{\tau > n\}} x_\tau \, d\mathbf{P}$$
$$= \mathbf{E}x_0 - \int_{\{\tau > n\}} x_n \, d\mathbf{P} + \int_{\{\tau > n\}} x_\tau \, d\mathbf{P}.$$

Hence by (1.37) and (1.38) we obtain the desired equality (1.36).

For the proof of Theorems 3—6 in full generality see **[49]**, Chapters V and VI, **[22]**, Chapter VII, and **[68]**.

CHAPTER II

OPTIMAL STOPPING OF MARKOV RANDOM SEQUENCES

§ 1. Statements of the problems. Excessive characterization of the value under the condition A^-

1. Suppose $X = (x_n, \mathscr{F}_n, \mathbf{P}_x)$, $n \in \boldsymbol{N}$, is a nonterminating Markov chain in the phase space[1] $(E, \mathscr{B})$. We let L denote the set of $\mathscr{B}$-measurable functions $g = g(x)$, $x \in E$, such that $-\infty < g(x) \leqq \infty$ and $\mathbf{E}_x g^-(x_n) < \infty$[2] for $n \in \boldsymbol{N}$ and $x \in E$. Let $L(A^-)$ and $L(A^+)$ be the collection of functions from L satisfying the additional conditions

$$A^-: \mathbf{E}_x \left[\sup_n g^-(x_n)\right] < \infty, \qquad x \in E, \tag{2.1}$$
$$A^+: \mathbf{E}_x \left[\sup_n g^+(x_n)\right] < \infty, \qquad x \in E,$$

respectively. We also let $L(A^-, A^+) = L(A^-) \cap L(A^+)$.

Let $\overline{\mathfrak{M}} = \{\tau\}$ be a class of Markov times (M.t.'s) $\tau = \tau(\omega)$ (relative to the system $\{\mathscr{F}_n\}$, $n \in \boldsymbol{N}$) with values in $\bar{\boldsymbol{N}} = \boldsymbol{N} \cup \{\infty\}$, and let $\mathfrak{M} \subseteq \overline{\mathfrak{M}}$ be the class of finite Markov times (stopping times, s.t.'s), i.e., of those $\tau \in \overline{\mathfrak{M}}$ for which $\mathbf{P}_x(\tau < \infty) = 1$ for all $x \in E$.

For each Markov time $\tau \in \overline{\mathfrak{M}}$ we let $\mathbf{E}_x\, g(x_\tau)$ denote the integral of the function $g(x_\tau)$ with respect to the measure $\mathbf{P}_x$, taken over the entire domain of definition of $g(x_\tau)$, i.e., we let

$$\mathbf{E}_x\, g(x_\tau) = \int_{(\tau<\infty)} g(x_\tau)\, d\mathbf{P}_x \qquad (= \mathbf{E}_x\, [g(x_\tau);\, \tau < \infty]).$$

Suppose that for $g \in L(A^-)$

$$s(x) = \sup_{\tau \in \mathfrak{M}} \mathbf{E}_x\, g(x_\tau) \tag{2.2}$$

and

$$\bar{s}(x) = \sup_{\tau \in \overline{\mathfrak{M}}} \tilde{\mathbf{E}}_x\, g(x_\tau), \tag{2.3}$$

where, by definition,

$$\tilde{\mathbf{E}}_x\, g(x_\tau) = \tilde{\mathbf{E}}_x\, [g(x_\tau);\, \tau < \infty] + \mathbf{E}_x \left[\limsup_n g(x_n);\, \tau = \infty\right].$$

[1] During the entire chapter we assume that $\mathscr{F} = \overline{\mathscr{F}}$, $\mathscr{F}_n = \overline{\mathscr{F}}_n$ and $\mathbf{P}_x = \overline{\mathbf{P}}_x$ for $n \in \boldsymbol{N}$ and $x \in E$.

[2] $g^- = -\min(g, 0)$, $g^+ = \max(g, 0)$.

By the condition A^-, the mathematical expectations in (2.2) and (2.3) are defined, and $\bar{s}(x) \geqq s(x) > -\infty$ for all $x \in E$. In the case of nonnegative functions $g \in L$ (and only in this case) we also let

$$\tilde{s}(x) = \sup_{\tau \in \mathfrak{M}} \mathbf{E}_x\, g(x_\tau).$$

It is clear that if $g \in L$ is nonnegative, then $\tilde{s}(x) \geqq \bar{s}(x) \geqq s(x) \geqq 0$.

DEFINITION 1. Each of the functions $s(x)$, $\bar{s}(x)$ and $\tilde{s}(x)$ is called a *value*. The stopping time $\tau_\varepsilon \in \mathfrak{M}$ is said to be *(ε,s)-optimal* if $s(x) - \varepsilon \leqq \mathbf{E}_x g(x_{\tau_\varepsilon})$ for all $x \in E$. The Markov times $\tau_\varepsilon \in \mathfrak{M}$ are said to be $(\varepsilon,\bar{s})$- and *$(\varepsilon,\tilde{s})$-optimal* if, respectively.

$$\bar{s}(x) - \varepsilon \leqq \mathbf{E}_x g(x_{\tau_\varepsilon}), \qquad x \in E,$$
$$\tilde{s}(x) - \varepsilon \leqq \tilde{\mathbf{E}}_x g(x_{\tau_\varepsilon}), \qquad x \in E.$$

Times which are $(0,s)$-, $(0,\bar{s})$-, or $(0,\tilde{s})$-optimal are simply called respectively s-, $\bar{s}$-, or *$\tilde{s}$-optimal.*

As will become clear below (see Theorem 1), every (ε,s)-optimal s.t. is at the same time $(\varepsilon,\bar{s})$-optimal as well as (in the case $g(x) \geqq 0$) $(\varepsilon,\tilde{s})$-optimal. Therefore, for brevity, we also call (ε,s)-optimal times ε-optimal s.t.'s. We shall call a 0-optimal s.t. an *optimal* s.t.

From the point of view of these definitions, a consideration of the value $\bar{s}(x)$ for *all* functions $g \in L(A^-)$ may not turn out to be meaningful, since if, say, $g(x) \equiv -1$ then by the given definition $\bar{s}(x) \equiv 0$. At the same time it is clear that a value which is thought of as the maximum possible mean reward in stopping at random moments of time cannot be larger than $\sup_{x \in E} g(x) = -1$.

It is often said that the stopping time assigns some *stopping rule.*

Our aim is to explain the structure of the values $s(x)$, $\bar{s}(x)$ and $\tilde{s}(x)$, and find conditions for existence of optimal and ε-optimal (in the sense indicated above) Markov times. In Theorem 1 we show that indeed $s(x) = \bar{s}(x)$ (and then, in the particular case of nonnegative $g \in L$, obviously $s(x) = \bar{s}(x) = \tilde{s}(x)$), where $s(x)$ is the smallest excessive majorant of the function $g(x)$. In Theorem 2 (§ 2) we give conditions for existence of ε-optimal ($\varepsilon \geqq 0$) Markov times and methods for finding them. The analogous problems for functions $g(x)$ from the classes $L(A^+)$ and $L(A^-,A^+)$ are considered in § 3.

2. DEFINITION 2. The $\mathscr{B}$-measurable function $f(x) \in L$ is said to be *excessive* if $Tf(x) \leqq f(x)$, $x \in E$, where $Tf(x) = T_1 f(x) = \mathbf{E}_x f(x_1)$.

An excessive function (e.f.) $f = f(x)$ is called an *excessive majorant* (e.m.) of the function $g(x)$ if $f(x) \geqq g(x)$, $x \in E$.

An excessive majorant $f(x)$ is called the *smallest excessive majorant*[3)] (s.e.m.) of the function $g(x)$ if $f(x)$ is less than or equal to every e.m. of the function $g(x)$.

Later[3)] (see Lemma 4) it will be shown that smallest e.m.'s of functions $g(x) \in L$ exist; moreover, a practical method is given for constructing them.

The usefulness of the concepts introduced above in optimal stopping problems for Markov chains is demonstrated in the following theorem.

THEOREM 1. *Suppose the function* $g \in L(A^-)$. *Then the following assertions are true. The value* $s(x)$ *is the smallest excessive majorant of the function* $g(x)$.

1) *The value* $s(x)$ *is the smallest excessive majorant of the function* $g(x)$.
2) $s(x) = \bar{s}(x)$.
3) $s(x) = \max\{g(x), Ts(x)\}$.
4) *If the function* $g(x)$ *is nonnegative, then* $s(x) = \bar{s}(x) = \tilde{s}(x)$.

For the proof of the theorem we need to examine in detail the properties of e.m.'s and s.e.m.'s.

Let $\mathscr{E}$ denote the set of excessive functions for the Markov chain X, and let $\mathscr{E}^+$ be the set of nonnegative e.f.'s from $\mathscr{E}$. Properties I—VI of excessive functions, given below, follow directly from their definition.

I. The function $f(x) \equiv C = \text{const}$ is excessive.

II. If $f, g \in \mathscr{E}$ and the constants $a, b \geqq 0$, then $af + bg \in \mathscr{E}$.

III. If $f \in \mathscr{E}$, then $T_t f(x) = \mathbf{E}_x f(x_t)$, $t \in \boldsymbol{N}$, is excessive, and $T_t f(x) \geqq T_{t+1} f(x)$.

IV. If $f_n \in \mathscr{E}$, $n = 1,2,\cdots$, and $f_{n+1} \geqq f_n$ for all n, then $f(x) = \lim_{n\to\infty} f_n(x) \in \mathscr{E}$.

V. If $f \in \mathscr{E}$, then for each $x \in E$ the system $\{f(x_n), \mathscr{F}_n, \mathbf{P}_x\}$, $n \in \boldsymbol{N}$, forms a (generalized) super-martingale:

$$\mathbf{E}_x f^-(x_n) < \infty, \quad \mathbf{E}_x[f(x_{n+1}) \mid \mathscr{F}_n] \leqq f(x_n), \quad n \in \boldsymbol{N}, \quad (\mathbf{P}_x\text{-a.s.}).$$

VI. If $f, g \in \mathscr{E}^+$, then the function $f \wedge g = \min(f, g) \in \mathscr{E}^+$. If $f \in \mathscr{E}$ and C is a constant, then $f^C = f \wedge C \in \mathscr{E}$.

The next important property follows from property V and Theorem 1.3:

VII. If $f \in \mathscr{E}$ and $\sup_n \mathbf{E}_x f^-(x_n) < \infty$, then the limit (finite or equal to $+\infty$) $\lim f(x_n)$ exists for each $x \in E$ with $\mathbf{P}_x$-probability one.

In the study of the properties of the values $s(x)$, $\bar{s}(x)$ and $\tilde{s}(x)$, a fundamental role is played by

LEMMA 1. *Suppose* $f \in \mathscr{E}$ *and satisfies the condition* A^-. *Suppose the Markov times* $\tau, \sigma \in \mathfrak{M}$, *with* $\tau \geqq \sigma$ *with* $\mathbf{P}_x$*-probability one for each* $x \in E$. *Then*

$$\tilde{\mathbf{E}}_x f(x_\sigma) \geqq \tilde{\mathbf{E}}_x f(x_\tau), \qquad x \in E, \tag{2.4}$$

or, more precisely

$$\int_{(\sigma<\infty)} f(x_\sigma) d\mathbf{P}_x + \int_{(\sigma=\infty)} \limsup_n f(x_n) d\mathbf{P}_x \geqq \int_{(\tau<\infty)} f(x_\tau)\, d\mathbf{P}_x + \int_{(\tau=\infty)} \limsup_n f(x_n)\, d\mathbf{P}_x. \tag{2.5}$$

In particular,

$$f(x) \geqq \int_{(\tau<\infty)} f(x_\tau)d\mathbf{P}_x + \int_{(\tau=\infty)} \limsup_n f(x_n)d\mathbf{P}_x. \tag{2.6}$$

PROOF. First of all we note that, by property VII,

$$\limsup_n f(x_n) = \lim_n f(x_n),$$

so that, indeed, we can replace the $\limsup_n$ in formulas (2.5) and (2.6) simply by $\lim_n$.

For the proof of (2.5), we first assume that the function $f(x) \leqq C < \infty$. Then for each $n \in \boldsymbol{N}$, by Theorem 1.5, for the times $\sigma_n = \sigma \wedge n$ and $\tau_n = \tau \wedge n$, we have

$$\begin{aligned}\int_\Omega f(x_{\sigma_n})d\mathbf{P}_x &= \int_{(\sigma<n)} f(x_\sigma)d\mathbf{P}_x + \int_{(\sigma\geqq n)} f(x_n)d\mathbf{P}_x \\ &\geqq \int_{(\tau<n)} f(x_\tau)d\mathbf{P}_x + \int_{(\tau\geqq n)} f(x_n)d\mathbf{P}_x = \int_\Omega f(x_{\tau_n})d\mathbf{P}_x,\end{aligned}$$

whence

$$\begin{aligned}&\int_{(\sigma<\infty)} f(x_\sigma)d\mathbf{P}_x + \int_{(\sigma=\infty)} f(x_n)d\mathbf{P}_x + \int_{(n\leqq\sigma<\infty)} [f(x_n) - f(x_\sigma)]\, d\mathbf{P}_x \\ &\geqq \int_{(\tau<\infty)} f(x_\tau)\, d\mathbf{P}_x + \int_{(\tau=\infty)} f(x_n)d\mathbf{P}_x + \int_{(n\leqq\tau<\infty)} [f(x_n) - f(x_\tau)]\, d\mathbf{P}_x,\end{aligned}$$

i.e.

$$\begin{aligned}\int_{(\sigma<\infty)} f(x_\sigma)d\mathbf{P}_x \geqq{}& \int_{(\tau<\infty)} f(x_\tau)d\mathbf{P}_x + \int_{(\tau=\infty)\setminus(\sigma=\infty)} f(x_n)d\mathbf{P}_x \\ &+ \int_{(n\leqq\tau<\infty)} [f(x_n) - f(x_\tau)]\, d\mathbf{P}_x - \int_{(n\leqq\sigma<\infty)} [f(x_n) - f(x_\sigma)]\, d\mathbf{P}_x.\end{aligned}$$

But

$$\lim_{n\to\infty}\left[\int_{(n\leqq\tau<\infty)} |f(x_n) - f(x_\tau)|\, d\mathbf{P}_x + \int_{(n\leqq\sigma<\infty)} |f(x_n) - f(x_\sigma)|\, d\mathbf{P}_x\right] = 0,$$

and therefore, by Fatou's Lemma ([**48**], Chapter II, § 7.2) and property VII,

$$\begin{aligned}\int_{(\sigma<\infty)} f(x_\sigma)d\mathbf{P}_x &\geqq \int_{(\tau=\infty)} f(x_\tau)d\mathbf{P}_x \\ &\qquad + \liminf_n \int_{(\tau=\infty)\setminus(\sigma=\infty)} f(x_n)d\mathbf{P}_x \\ &\geqq \int_{(\tau<\infty)} f(x_\tau)d\mathbf{P}_x + \int_{(\tau=\infty)\setminus(\sigma=\infty)} \liminf_n f(x_n)d\mathbf{P}_x \\ &= \int_{(\tau<\infty)} f(x_\tau)d\mathbf{P}_x + \int_{(\tau=\infty)\setminus(\sigma=\infty)} \lim_{n\to\infty} f(x_n)d\mathbf{P}_x,\end{aligned}$$

which (in the case of functions $f(x) \leqq C$) is equivalent to (2.5).

In the general case, with each function $f \in \mathscr{E}$ we associate the (excessive) functions $f^m = \min(f, m)$, $m \in \boldsymbol{N}$. Then $f^m(x) \uparrow f(x)$ as $m \to \infty$, and

$$\int_{(\sigma<\infty)} f(x_\sigma)d\mathbf{P}_x \geqq \int_{(\sigma<\infty)} f^m(x_\sigma)d\mathbf{P}_x$$
$$\geqq \int_{(\tau<\infty)} f^m(x_\tau)d\mathbf{P}_x + \int_{(\tau=\infty)\setminus(\sigma=\infty)} \lim_{n\to\infty} f^m(x_n)d\mathbf{P}_x, \tag{2.7}$$

whence, passing to the limit as $m\to\infty$ in the right side of (2.7), we obtain the desired inequality (2.5) if we make use of the relationship

$$\lim_{m\to\infty}\left(\lim_{n\to\infty} f^m(x_n)\right) = \lim_{n\to\infty} f(x_n) \qquad (\mathbf{P}_x\text{-a.s., } x\in E),$$

whose validity is easily established.

Indeed, according to Theorem I.3, for $x\in E$ the limit $\lim_{n\to\infty} f(x_n)$ is either finite or equal to $+\infty$, with $\mathbf{P}_x$-probability one. If $\bar{f} = \lim_{n\to\infty} f(x_n) < \infty$,then for sufficiently large m

$$\lim_{n\to\infty} \min\,(m, f(x_n)) = \bar{f},$$

and thus the desired relationship is established. Now if $\bar{f} = \lim_{n\to\infty} f(x_n) = +\infty$, then

$$\lim_{n\to\infty} \min\,(m, f(x_n) = m$$

and $\lim_{m\to\infty} m = +\infty$, which is what we wanted to prove.

From property (2.6) we can draw the following conclusion: if in (2.3) the function $g(x)\in L(A^-)$ and is excessive, then $\bar{s}(x)\equiv g(x)$ and the "best strategy" is to "stop" immediately. Indeed, according to (2.6),

$$\bar{\mathbf{E}}_x g(x_\tau) \leqq \mathbf{E}_x g(x_0) = g(x)$$

and consequently $\bar{s}(x)\equiv g(x)$.

The now-proved Lemma 1 allows us to establish the following important assertion.

Lemma 2. *If the function $f(x)$ is excessive and satisfies condition A^- and*[4] $\sigma_A = \inf\{n\colon x_n\in A\}$, $A\in\mathscr{B}$, *then the function*

$$f_A(x) = \mathbf{E}_x f(x_{\sigma_A}) \tag{2.8}$$

is also excessive.

Proof. Let $\sigma = \min\{n\geqq 1\colon x_n\in A\}$. Obviously σ is a M.t. and $\sigma\geqq\sigma_A$. Then

$$Tf_A(x) = \mathbf{E}_x f_A(x_1) = \mathbf{E}_x\mathbf{E}_{x_1} f(x_{\sigma_A}).$$

[4] Recall that σ_A is set equal to ∞ if the set $\{n\colon x_n\in A\}$ is empty.

By (1.21)

$$\mathbf{E}_x \mathbf{E}_{x_1} f(x_{\sigma_A}) = \mathbf{E}_x \theta_1 f(x_{\sigma_A}) = \mathbf{E}_x f(\theta_1 x_{\sigma_A}).$$

But it is known ([25], Chapter 4, property 4.1.E) that if

$$\sigma_A^s = \inf \{n \geqq s: x_n \in A\} \qquad (\sigma_A^0 = \sigma_A),$$

then $\theta_t\, x_{\sigma_A^s} = x_{\sigma_A^{s+t}}$. Therefore $\theta_1\, x_{\sigma_A} x_{\sigma_A^1} = x_\sigma$ and hence

$$Tf_A(x) = \mathbf{E}_x f(\theta_1 x_{\sigma_A}) = \mathbf{E}_x f(x_\sigma) \leqq \mathbf{E}_x f(x_{\sigma_A}) = f_A(x).$$

3. If the function $f = f(x)$ is an excessive majorant of the function $g \in L(A^-)$, then obviously $f \in L(A^-)$ and

$$f(x) \geqq \max \{g(x),\ Tf(x)\}. \tag{2.9}$$

If f is the *smallest* e.m. of the function g, then we have equality in (2.9). Indeed, we have

LEMMA 3. *If $\upsilon(x)$ is the s.e.m. of the function $g(x)$, then*

$$\upsilon(x) = \max \{g(x),\ T\upsilon(x)\}. \tag{2.10}$$

PROOF. From (2.9), $\upsilon(x) \geqq \max\{g(x),\ T\upsilon(x)\}$. The function $\upsilon_1(x) = \max \{g(x), T\upsilon(x)\}$ is an e.m. of the function $g(x)$ since $\upsilon_1(x) \geqq g(x)$ and

$$T\upsilon_1(x) \leqq T\upsilon(x) \leqq \max \{g(x), T\upsilon(x)\} = \upsilon_1(x).$$

But since $\upsilon(x)$ is the s.e.m., $\upsilon_1(x) \geqq \upsilon(x)$ and thus $\upsilon_1(x) = \upsilon(x)$.

4. Let $g(x) \in L(A^-)$. Does the smallest excessive majorant of the function $g(x)$ exist? A positive answer to this question is contained in the following lemma, which gives, in particular, a method of practical importance for finding the s.e.m.

LEMMA 4. *Suppose $g(x) \in L$. Let*

$$Qg(x) = \max\{g(x), Tg(x)\} \tag{2.11}$$

and $\upsilon(x) = \lim_{N\to\infty} Q^N g(x)$, where Q^N is *the Nth power of the operator Q. Then $\upsilon(x)$ is the s.e.m. of the function $g(x)$.*

PROOF. First we note that $Q^{N+1}g(x) \geqq Q^N g(x)$; therefore the limit $\lim_{N\to\infty} Q^N g(x)$ exists. It is also clear that

$$\upsilon(x) = \lim_{N\to\infty} Q^N g(x) \geqq g(x).$$

Now we verify that the function $\upsilon(x)$ is excessive. By the inequality $Q^N g(x) \geqq -\bar{g}(x)$, where $\mathbf{E}_x g^-(x_1) < \infty$, the inequality $Q^N g(x) \geqq TQ^{N-1}(x)$ and the Lebesgue monotone convergence theorem ([48], Chapter II, § 72), we have

$$v(x) = \lim_{N\to\infty} Q^N g(x) \geqq \lim_{N\to\infty} TQ^{N-1} g(x)$$
$$= T\left(\lim_{N\to\infty} Q^{N-1} g\right)(x) = Tv(x).$$

Consequently $v(x)$ is an e.m. of the function $g(x)$.

Suppose $f(x)$ is also an e.m. of the function $g(x)$. Then $f(x) \geqq Tf(x)$ and thus

$$Qf(x) = \max\{f(x),\ Tf(x)\} = f(x), \qquad f(x) = Q^N f(x) \geqq Q^N g(x),$$

whence $f(x) \geqq v(x)$, i.e. $v(x)$ is the smallest e.m. of the function $g(x)$.

REMARK 1. If we set

$$\tilde{Q}g(x) = \sup\,\{g(x),\ Tg(x),\ T_2 g(x), \ldots\}, \tag{2.12}$$

then we can show analogously that $\tilde{v}(x) = \lim_{N\to\infty} \tilde{Q}^N g(x)$ is also the smallest e.m. of the function $g(x)$ and consequently coincides with $v(x)$.

REMARK 2. Let $g \in L$ and $g^b(x) = \min\,(b, g(x))$, where $b \geqq 0$. Then the s.e.m. of the function $g(x)$ is the function

$$v(x) = \lim_{b\to\infty} \lim_{N\to\infty} Q^N g^b(x) = \lim_{N\to\infty} \lim_{b\to\infty} Q^N g^b(x).$$

Indeed, since $\lim_{b\to\infty} Q^N g^b(x) = Q^N g(x)$, then, as already proved,

$$v(x) = \lim_{N\to\infty} Q^N g(x) = \lim_{N\to\infty} \left(\lim_{b\to\infty} Q^N g^b(x)\right).$$

Moreover

$$v^b(x) = \lim_{N\to\infty} Q^N g^b(x) \leqq \lim_{N\to\infty} Q^N g(x) = v(x)$$

and consequently $\lim_{b\to\infty} v^b(x) \leqq v(x)$. But $\lim_{b\to\infty} v^b(x)$ is an excessive function (by property IV) and $\lim_{b\to\infty} v^b(x) \geqq g(x)$. Therefore

$$v(x) = \lim_{b\to\infty} v^b(x) = \lim_{b\to\infty} \lim_{N\to\infty} Q^N g^b(x).$$

REMARK 3. By (2.11) and the definition of the operators Q^N, $N \geqq 1$,

$$Q^N g(x) = \max\,\{Q^{N-1} g(x),\ TQ^{N-1} g(x)\}, \qquad Q^0 g(x) = g(x).$$

It is valuable to note that the sequence $\{Q^N g(x),\ N \geqq 1\}$ also satisfies the (simpler) recursion equations

$$Q^N g(x) = \max\{g(x),\ TQ^{N-1} g(x)\}, \qquad Q^0 g(x) = g(x).$$

The proof is easily carried out by induction. We shall show only that

$$Q^2 g(x) = \max\{g(x),\ TQg(x)\}.$$

Indeed, since $T[\max(g,Tg)](x) \geqq Tg(x)$, we have

$$\begin{aligned} Q^2g(x) &= \max\{Qg(x), TQg(x)\} \\ &= \max\{\max[g(x), Tg(x)], T[\max(g, Tg)](x)\} \\ &= \max\{g(x), T[\max(g, Tg)](x)\} = \max\{g(x), TQg(x)\}. \end{aligned}$$

5. LEMMA 5. *Suppose the excessive function $f(x)$ is such that $f(x) < \infty$, $x \in E$, and*

$$f(x) = \max\{g(x), Tf(x)\}, \qquad x \in E, \tag{2.13}$$

where $g \in L$. Let $\Gamma_\varepsilon = \{x: f(x) \leqq g(x) + \varepsilon\}$, $\varepsilon \geqq 0$, and $\tau_\varepsilon = \inf\{n: x_n \in \Gamma_\varepsilon\}$, $n \in \boldsymbol{N}$. Then for all $N \in \boldsymbol{N}$ and $x \in E$

$$f(x) = \int_{(\tau_\varepsilon \leqq N)} f(x_{\tau_\varepsilon}) d\mathbf{P}_x + \int_{(\tau_\varepsilon > N)} f(x_N) d\mathbf{P}_x. \tag{2.14}$$

PROOF. Since $\mathbf{E}_x f(x_k(\omega))$, $k \in \boldsymbol{N}$, is defined, we have

$$\begin{aligned} f(x) &= \int_{(\tau_\varepsilon \geqq 0)} f(x_0(\omega))\, d\mathbf{P}_x \\ &= \int_{(\tau_\varepsilon = 0)} f(x_0(\omega)) d\mathbf{P}_x + \int_{(\tau_\varepsilon > 0)} f(x_0(\omega)) d\mathbf{P}_x. \end{aligned} \tag{2.15}$$

But $f(x_k(\omega)) > g(x_k(\omega))$ on the set $\{\tau_\varepsilon > k\}$, and consequently $f(x_k(\omega)) = Tf(x_k(\omega))$. Therefore from (2.15)

$$\begin{aligned} f(x) &= \int_{(\tau_\varepsilon = 0)} f(x_0) d\mathbf{P}_x + \int_{(\tau_\varepsilon > 0)} f(x_1) d\mathbf{P}_x \\ &= \int_{(0 \leqq \tau_\varepsilon \leqq 1)} f(x_{\tau_\varepsilon}) d\mathbf{P}_x + \int_{(\tau_\varepsilon > 1)} f(x_1) d\mathbf{P}_x = \dots \\ &= \int_{(0 \leqq \tau_\varepsilon \leqq N)} f(x_{\tau_\varepsilon}) d\mathbf{P}_x + \int_{(\tau_\varepsilon > N)} f(x_N) d\mathbf{P}_x, \end{aligned}$$

which proves (2.13).

REMARK. The smallest excessive majorant $v(x)$ of the function $g(x)$ satisfies equation (2.13). However not every function $f(x)$ which is a solution of equation (2.13) is the s.e.m. of the function $g(x)$. Indeed, if $g(x) \leqq C < \infty$, then every constant greater than C satisfies (2.13).

6. An essential role in the proof of Theorem 1 is played by

LEMMA 6. *Suppose the function $g \in L$ is such that for all $x \in E$ the following condition is satisfied*:

$$A^+: \mathbf{E}_x\left[\sup_n g^+(x_n)\right] < \infty, \qquad x \in E, \tag{2.16}$$

i.e. $g \in L(A^+)$. Suppose $v(x)$ is the s.e.m. of the function $g(x)$. Then

$$\limsup_n g(x_n) = \limsup_n v(x_n) \qquad (\mathbf{P}_x\text{-}a.s., x \in E). \tag{2.17}$$

PROOF. Since $\upsilon(x) \geqq g(x)$,

$$\limsup_n \upsilon(x_n) \geqq \limsup_n g(x_n).$$

We let

$$\varphi_n(\omega) = \mathbf{E}_x\{\sup_{j \geqq n} g(x_j) \mid \mathscr{F}_n\}.$$

Making use of the Markov property, we have $\varphi_n(\omega) = \varphi(x_n(\omega))$ ($\mathbf{P}_x$-a.s., $x \in E$), where $\varphi(x) = \mathbf{E}_x\{\sup_{j \geqq 0} g(x_j)\}$. We note that the function $\varphi(x)$ is excessive. Indeed, $\mathbf{E}_x\varphi^-(x_n) < \infty$, $n \in \boldsymbol{N}$, $x \in E$, and

$$\begin{aligned} T\varphi(x) &= \mathbf{E}_x\varphi(x_1) = \mathbf{E}_x\Big\{\mathbf{E}_{x_1}\Big[\sup_{j \geqq 0} g(x_j)\Big]\Big\} \\ &= \mathbf{E}_x\{\sup_{j \geqq 1} g(x_j)\} \leqq \mathbf{E}_x\{\sup_{j \geqq 0} g(x_j)\} = \varphi(x). \end{aligned}$$

It is also clear that $\varphi(x) \geqq g(x)$, whence $\varphi(x) \geqq \upsilon(x)$ and

$$\varphi_n(\omega) = \varphi(x_n(\omega)) \geqq \upsilon(x_n(\omega)) \qquad (\mathbf{P}_x\text{-a.s.}).$$

Suppose m is fixed and $n \geqq m$ ($m, n \in \boldsymbol{N}$). Then

$$\mathbf{E}_x\Big[\sup_{j \geqq m} g(x_j) \mid \mathscr{F}_n\Big] \geqq \mathbf{E}_x\Big[\sup_{j \geqq n} g(x_j) \mid \mathscr{F}_n\Big] \geqq \upsilon(x_n) \qquad (\mathbf{P}_x\text{-a.s., } x \in E)$$

and

$$\limsup_n \tilde{\mathbf{E}}_x\Big[\sup_{j \geqq m} g(x_j) \mid \mathscr{F}_n\Big] \geqq F \limsup_n \upsilon(x_n) \qquad (\mathbf{P}_x\text{-a.s., } x \in E). \qquad (2.18)$$

For all $m \in \boldsymbol{N}$ and $x \in E$ the sequence

$$\{\mathbf{E}_x[\psi_m \mid \mathscr{F}_n], \mathscr{F}_n, \mathbf{P}_x\} \qquad n \geqq m,$$

where $\psi_m = \sup_{j \geqq m} g(x_j)$, forms a generalized martingale by the condition A^+.

According to Theorem 1.4, the limit $\lim_{n \to \infty} \mathbf{E}_x[\psi_m \mid \mathscr{F}_n]$ exists with $\mathbf{P}_x$-probability one and

$$\lim_{n \to \infty} \mathrm{E}_x[\psi_m \mid \mathscr{F}_n] \leqq \mathbf{E}_x[\psi_m \mid \mathscr{F}_\infty],$$

where $\mathscr{F}_\infty = \sigma(\bigcup_{n \in N} \mathscr{F}_n)$. Since the random variable ψ_m is $\mathscr{F}_\infty$-measurable,

$$\mathbf{E}_x[\psi_m \mid \mathscr{F}_\infty] = \psi_m \qquad (\mathbf{P}_x\text{-a.s.}).$$

Therefore

$$\begin{aligned} &\limsup_n \mathbf{E}_x\Big[\sup_{j \geqq m} g(x_j) \mid \mathscr{F}_n\Big] \\ &\quad = \limsup_n \mathbf{E}_x[\psi_m \mid \mathscr{F}_n] = \lim_n \mathbf{E}_x[\psi_m \mid \mathscr{F}_n] \leqq \psi_m \end{aligned}$$

and by (2.18)

$$\psi_m \leqq \limsup_n \upsilon(x_n) \qquad (\mathbf{P}_x\text{-a.s.}).$$

Hence

$$\limsup_n g(x_n) = \inf_m \Big[\sup_{j \geqq m} g(x_j)\Big] = \inf_m \psi_m \geqq \limsup_n v(x_n)$$

$\mathbf{P}_x$-a.s. for each $x \in E$, which proves the lemma.

REMARK. If $\sup_n \mathbf{E}_x v^-(x_n) < \infty$ for $x \in E$, then by property VII,

$$\limsup_n v(x_n) = \lim_n v(x_n) \qquad (\mathbf{P}_x\text{-a.s., } x \in E).$$

7. The sequence $Q^N g(x)$ constructed for the function $g \in L(A^-)$ in Lemma 4, which is monotone increasing (more precisely, not decreasing) as $N \to \infty$ and approaches $v(x)$, is the s.e.m. of the function $g(x)$.

Useful also in many cases is the sequence of functions constructed below, which (as will be shown in Lemma 8) is monotone decreasing (more precisely, not increasing) for functions from $L(A^-, A^+)$ and converges to $v(x)$, the s.e.m.

First we introduce some notation. Let $g \in L$. With this function we associate the operator

$$Gf(x) = \max\{g(x), Tf(x)\}, \qquad f \in L, \tag{2.19}$$

and use $G^N f(x)$ to denote the Nth power of the operator $Gf(x)$; $G^0 f(x) = f(x)$. (If $f = g$, then $Gg(x) = Qg(x)$, where Q is the operator defined in (2.11). If $f(x) = v(x)$ is the s.e.m. of the function $g(x)$, then (see Lemma 3) $Gv(x) = v(x)$.)

We prove the following assertion.[5)]

LEMMA 7. *Let* $g \in L(A^+)$ *and* $\varphi(x) = \mathbf{E}_x[\sup_n g(x_n)]$. *Then*

$$G^{N+1}\varphi(x) \leqq G^N \varphi(x), \qquad N = \boldsymbol{N},$$

and the function $\tilde{v}(x) = \lim_{N\to\infty} G^N \varphi(x)$ *satisfies the equation*

$$\tilde{v}(x) = \max\{g(x), T\tilde{v}(x)\}. \tag{2.20}$$

PROOF. The inequality $G^{N+1}\varphi(x) \leqq G^N\varphi(x)$ can be verified by induction. We shall show only that $G\varphi(x) \leqq \varphi(x)$. Indeed,

$$\begin{aligned} G\varphi(x) &= \max\{g(x), T\varphi(x)\} \\ &= \max\Big\{g(x), \mathbf{E}_x\Big[\sup_{j\geqq 1} g(x_j)\Big]\Big\} \\ &\leqq \mathbf{E}_x\Big\{\max\Big[g(x), \sup_{j\geqq 1} g(x_j)\Big]\Big\} = \mathbf{E}_x\Big\{\sup_{j\geqq 0} g(x_j)\Big\} = \varphi(x). \end{aligned}$$

Further, since

$$\varphi(x) \geqq G^N\varphi(x) \downarrow v(x), \qquad \text{as } N \to \infty,$$
$$\boldsymbol{T}\varphi(x) < \infty, \qquad x \in E,$$

5) Lemmas 7, 8 and 9 will not be used directly in the proof of Theorem 1.

on passing to the limit as $N \to \infty$ in the equality

$$G^N\varphi(x) = \max\{g(x), TG^{N-1}\varphi(x)\}$$

and using the Lebesgue monotone convergence theorem, we obtain that $v(x)$ satisfies equation (2.20).

Thus the function $\tilde{v}(x)$, which is an e.m. of the function $g \in L(A^+)$, need not be the s.e.m. (see the example in § 3). However if $g \in L(A^+, A^-)$, then the function $\tilde{v}(x)$ coincides with $v(x)$, the s.e.m. of $g(x)$. To prove this result we establish the following preliminary assertion.

LEMMA 8. *Let $g \in L(A^+)$ and $v(x) = \lim_{N\to\infty} G^N\varphi(x)$. Then $\mathbf{P}_x$-a.s., $x \in E$,*

$$\limsup_n \tilde{v}(x_n) = \limsup_n g(x). \tag{2.21}$$

PROOF. The inequality $\limsup_n v(x_n) \geqq \limsup_n g(x_n)$ is obvious. On the other hand, for all $x \in E$, $n \in \boldsymbol{N}$ and $m \leqq n$

$$\begin{aligned}\tilde{v}(x_n) &\leqq G^N\varphi(x_n) \leqq \varphi(x_n) = \mathbf{E}_{x_n}\Big[\sup_{i\geqq 0} g(x_j)\Big]\\ &= \mathbf{E}_x\Big[\sup_{j\geqq n} g(x_j) \mid \mathscr{F}_n\Big] \leqq \mathbf{E}_x\Big[\sup_{j\geqq m} g(x_j) \mid \mathscr{F}_n\Big]\\ &\quad (\mathbf{P}_x\text{-a.s., } x \in E).\end{aligned} \tag{2.22}$$

From (2.22) (as in Lemma 6) it follows that

$$\limsup_n \tilde{v}(x_n) = \sup_{j\geqq m} g(x_j),$$

and consequently

$$\limsup_n \tilde{v}(x_n) \leqq \inf_m \Big[\sup_{j\geqq m} g(x_j)\Big] = \limsup_n g(x_n),$$

which proves the lemma.

LEMMA 9. *Suppose $g \in L(A^+, A^-)$, $v(x)$ is its s.e.m.*[6] *and $\tilde{v}(x) = \lim_{N\to\infty} G^N\varphi(x)$. Then $v(x) = \tilde{v}(x)$ for all $x \in E$.*

PROOF. We define the time $\tilde{\tau}_\varepsilon = \inf\{n : \tilde{v}(x_n) \leqq g(x_n) + \varepsilon\}$, where $\varepsilon > 0$. By (2.21), $\mathbf{P}_x(\tilde{\tau}_\varepsilon < \infty) = 1$ for all $x \in E$. Indeed, let $A = \{\omega : \tilde{\tau}_\varepsilon(\omega) = \infty\}$. For $\omega \in A$ and all $n \in \boldsymbol{N}$ we have $\tilde{\tau}(x_n(\omega)) > g(x_n(\omega)) + \varepsilon$, and thus

$$\limsup_n \tilde{v}(x_n(\omega)) > \limsup_n g(x_n(\omega)). \tag{2.23}$$

By comparing (2.21) and (2.23) we now obtain $\mathbf{P}_x(A) = 0$, $x \in E$, since $\mathbf{P}_x(\limsup_n g(x_n) = \pm\infty) = 0$ for all $x \in E$.

[6] By Lemma 4, the s.e.m. $v(x)$ of the function $g \in L(A^-)$ exists.

We now apply Lemma 5, with $f(x) = \tilde{v}(x)$. Then, by (2.14), for all $N \in \mathbf{N}$

$$\tilde{v}(x) = \int_{(\bar{\tau}_\varepsilon \leqq N)} \tilde{v}(x_{\bar{\tau}_\varepsilon}) d\mathbf{P}_x + \int_{(\bar{\tau}_\varepsilon > N)} \tilde{v}(x_N) d\mathbf{P}_x. \tag{2.24}$$

We note that since $\mathbf{P}_x(\bar{\tau}_\varepsilon < \infty) = 1$, $x \in E$, then by (2.21)

$$\begin{aligned}\limsup_N \int_{(\bar{\tau}_\varepsilon > N)} \tilde{v}(x_N) d\mathbf{P}_x & \leqq \limsup_N \int_{(\bar{\tau}_\varepsilon > N)} \mathbf{E}_{x_N}\Big[\sup_{n \geqq 0} g^+(x_n)\Big] d\mathbf{P}_x \\ &= \limsup_N \int_{(\bar{\tau}_\varepsilon > N)} \sup_{n \geqq N} g^+(x_n)\, d\mathbf{P}_x \\ &\leqq \limsup_N \int_{(\bar{\tau}_\varepsilon > N)} \sup_{n \geqq 0} g^+(x_n) d\mathbf{P}_x = 0.\end{aligned}$$

Analogously

$$\begin{aligned}\liminf_N \int_{(\bar{\tau}_\varepsilon > N)} \tilde{v}(x_N) d\mathbf{P}_x \\ \geqq - \liminf_N \int_{(\bar{\tau}_\varepsilon > N)} \mathbf{E}_{x_N}\Big[\sup_{n \geqq 0} g^-(x_n)\Big] d\mathbf{P}_x = 0.\end{aligned}$$

Therefore, passing to the limit as $N \to \infty$ in (2.24), we obtain

$$\tilde{v}(x) = \mathbf{E}_x \tilde{v}(x_{\bar{\tau}_\varepsilon}). \tag{2.25}$$

But $\tilde{v}(x_{\bar{\tau}_\varepsilon}) \leqq g(x_{\bar{\tau}_\varepsilon}) + \varepsilon \leqq v(x_{\bar{\tau}_\varepsilon}) + \varepsilon$, and therefore from (2.25) and (2.6)

$$\begin{aligned}\tilde{v}(x) = \mathbf{E}_x \tilde{v}(x_{\bar{\tau}_\varepsilon}) &\leqq \mathbf{E}_x g(x_{\bar{\tau}_\varepsilon}) \\ &+ \varepsilon \leqq \mathbf{E}_x v(x_{\bar{\tau}_\varepsilon}) + \varepsilon \leqq v(x) + \varepsilon\end{aligned} \tag{2.26}$$

Since $\varepsilon > 0$ was arbitrary, it now follows that $\tilde{v}(x) \leqq v(x)$, and hence $\tilde{v}(x) = v(x)$, since the inequality $\tilde{v}(x) \geqq v(x)$ is obvious.

8. Proof of Theorem 1. Let $v(x)$ be the s.e.m. of the function $g \in L(A^-)$. Then by Lemma 1, for $\tau \in \overline{\mathfrak{M}}$

$$\begin{aligned}v(x) &\geqq \int_{(\tau < \infty)} v(x_\tau) d\mathbf{P}_x + \int_{(\tau = \infty)} \limsup_n v(x_n)\, d\mathbf{P}_x \\ &\geqq \int_{(\tau > \infty)} g(x_\tau) d\mathbf{P}_x + \int_{(\tau = \infty)} \limsup_n g(x_n) d\mathbf{P}_x.\end{aligned}$$

Consequently

$$v(x) \geqq \sup_{\tau \in \overline{\mathfrak{M}}} \bar{\mathbf{E}}_x g(x_\tau) = \bar{s}(x) \geqq s(x). \tag{2.27}$$

For a complete proof of the theorem it is sufficient to show that $v(x) \leqq s(x)$.

First suppose that the function $g(x)$ also satisfies the condition A^+: $\mathbf{E}_x [\sup_n g^+(x_n)] < \infty$ for all $x \in E$. Then, by Lemma 6,

$$\limsup_n v(x_n) = \limsup_n g(x_n),$$

whence it follows that the time

$$\tau_\varepsilon = \inf\{n: v(x_n) \leqq g(x_n) + \varepsilon\}, \qquad \varepsilon > 0, \tag{2.28}$$

(as in Lemma 9) is a stopping time, $\tau_\varepsilon \in \mathfrak{M}$, where (according to Lemma 6) $v(x) = \mathbf{E}_x v(x_{\tau_\varepsilon})$.

Thus (cf. (2.26))

$$v(x) = \mathbf{E}_x v(x_{\tau_\varepsilon}) \leqq \mathbf{E}_x g(x_{\tau_\varepsilon}) + \varepsilon \leqq s(x) + \varepsilon. \tag{2.29}$$

From (2.29), since $\varepsilon > 0$ was arbitrary, $v(x) \leqq s(x)$, which together with (2.27) proves Theorem 1 under the condition A^+, which is obviously satisfied if the function $g(x) \leqq C < \infty$.

For the proof of the inequality $v(x) \leqq s(x)$ in the general case we let $g^b(x) = \min(b, g(x))$, $b \geqq 0$, and we let $v^b(x)$ be the s.e.m. of the function $g^b(x)$ and $s^b(x) = \sup_{\tau \in \mathfrak{M}} \mathbf{E}_x g^b(x_\tau)$. Then by what was proved above

$$s(x) \geqq s^b(x) = \sup_{\tau \in \mathfrak{M}} \mathbf{E}_x g^b(x_\tau) \geqq v^b(x).$$

The sequence $\{v^b(x),\ b \geqq 0\}$ does not decrease. Let $\bar{v}(x) = \lim_{b\to\infty} v^b(x)$. We shall prove that indeed $\bar{v}(x) = v(x)$. We have

$$T\bar{v}(x) = T\left(\lim_{b\to\infty} v^b\right)(x) = \lim_{b\to\infty} Tv^b(x) \leqq \lim_{b\to\infty} v^b(x) = \bar{v}(x),$$

i.e. the function $\bar{v}(x)$ is excessive. Since $g^b(x) \uparrow g(x)$ and $v^b(x) \geqq g^b(x)$, it follows that $\bar{v}(x) \geqq g(x)$. Consequently $\bar{v}(x)$ is an e.m. of the function $g(x)$. It remains only to show that this e.m. is the smallest one. Let $f(x)$ be an e.m. of the function $g(x)$. Then $f(x) \geqq g^b(x)$ and $f(x) \geqq v^b(x)$, whence $f(x) \geqq \bar{v}(x)$. Consequently $s(x) \geqq \bar{v}(x) = v(x)$, which is what we wanted to prove.

The recursion equation $s(x) = \max\{g(x), Ts(x)\}$ follows in an obvious fashion from Lemma 3 and the equality $s(x) = v(x)$.

9. COROLLARY 1. *Suppose the time $\tau^* \in \overline{\mathfrak{M}}$ is such that its associated "reward" $\bar{f}(x) = \mathbf{E}_x g(x_{\tau^*})$, $g(x) \geqq 0$ $(\tilde{f}(x) = \tilde{\mathbf{E}}_x\, g(x_{\tau^*}))$ is an excessive function and $\bar{f}(x) \geqq g(x)$ $(\tilde{f}(x) \geqq g(x))$. Since $\bar{s}(x)$ $(\tilde{s}(x))$ is the s.e.m. of $g(x)$, then*[7] *$\bar{f}(x) = \bar{s}(x)$ $(\tilde{f}(x) = \tilde{s}(x))$ and the time $\tau^* \in \overline{\mathfrak{M}}$ is $(0,\bar{s})$-optimal (respectively $(0,\tilde{s})$-optimal).*

We note that, generally speaking, a $(0,\bar{s})$-optimal time need not be $(0,\tilde{s})$-optimal.

COROLLARY 2. *Let $\hat{\mathfrak{M}} \subseteq \mathfrak{M}$ be the class of Markov stopping times $\sigma_A = \inf\{n: x_n \in A\}$, where $A \in \mathscr{B}$. Then if $g \in L(A^-)$, we have $\bar{s}(x) = s(x) = \hat{s}(x)$, where*

---

[7] Recall that our consideration of $\bar{s}(x)$ assumes nonnegativity of the function $g(x)$.

$$\hat{s}(x) = \sup_{\tau \in \hat{\mathfrak{M}}} \mathbf{E}_x g(x_\tau). \tag{2.30}$$

This result, which follows directly from the proof of Theorem 1, emphasizes the importance of consideration of the class $\hat{\mathfrak{M}}$ of (Markov) times of first entry of the Borel sets $A \in \mathscr{B}$ in problems on optimal stopping of Markov chains. But this of course does not mean that the optimal (in one of the three given senses) time, if it exists, must necessarily be the time of first entry of some set $A \in \mathscr{B}$.

10. We give an example which illustrates Corollary 1.

EXAMPLE. Let $\{\xi_n\}$, $n \in \boldsymbol{N}$, be a sequence of independent identically distributed random variables taking on the two values $+1$ and -1 with probabilities $\mathbf{P}(\xi_i = +1) = p$ and $\mathbf{P}(\xi_i = -1) = q = 1-p$. Let $x_0 = x$ and $x_n = x + \xi_1 + \ldots + \xi_n$, where $x \in E = \{0, \pm 1, \pm 2, \ldots\}$. Then the process $X = \{x_n, \mathscr{F}_n, \mathbf{P}_x\}$, $n \in \boldsymbol{N}$, forms a Markov chain with values in E, where $\mathscr{F}_n = \sigma\{\omega: x_0, \ldots, x_n\}$ and $\mathbf{P}_x$ is the probability distribution on sets from $\mathscr{F}_\infty = \sigma(\bigcup_{n \in \boldsymbol{N}} \mathscr{F}_n)$, corresponding to an initial state x and the naturally induced random variables $\{\xi_n\}$, $n \in \boldsymbol{N}$.

Suppose the function $g(x) = \max\{0, x\}$. It is easy to see that in the case $p \geqq q$ the time $\tau^* \equiv \infty$ is $(0,\bar{s})$-optimal, where $\bar{s}(x) \equiv \infty$ for all $x \in E$. The case $p < q$ is of greater interest. We shall show that in this situation there exists a $(0,\bar{s})$-optimal time τ^* which is not a stopping time (i.e. $\tau^* \notin \mathfrak{M}$).

We define the Markov times $\tau_\gamma = \inf\{n: x_n \in [\gamma, \infty)\}$, where $\gamma \in E$. It is easy to show that the probability $p_\gamma(x)$ of entry of the set $\Gamma_\gamma = [\gamma, \infty)$, $x_0 = x$, for various $x \in E$ is given by the equalities

$$p_\gamma(x) = \begin{cases} (p/q)^{\gamma - x}, & x \leqq \gamma, \\ 1, & x \geqq \gamma. \end{cases}$$

Therefore

$$f_\gamma(x) = \mathbf{E}_x g(x_{\tau_\gamma}) = \begin{cases} \gamma(p/q)^{\gamma - x}, & x \leqq \gamma, \\ x, & x \geqq \gamma. \end{cases}$$

Let $f^*(x) = \sup_{\gamma \in E} f_\gamma(x)$. Then $f^*(x) = f_{\gamma^*}(x)$, where γ^* is the maximum point of the function $\gamma(p/q)^\gamma$ on the set E, and

$$f^*(x) = \begin{cases} \gamma^*(p/q)^{\gamma^* - x}, & x \leqq \gamma^*, \\ x, & x \geqq \gamma^*. \end{cases}$$

It is easy to see that $f^*(x) \geqq g(x)$ and $f^*(x) \geqq Tf^*(x)$ for all $x \in E$. Hence it follows that the time $\tau^* = \tau_{\gamma^*}$ is $(0,\bar{s})$-optimal:

$$\bar{s}(x) = \sup_{\tau \in \bar{\mathfrak{M}}} \mathbf{E}_x g(x_\tau) = \mathbf{E}_x g(x_{\tau^*}).$$

Finally, this time is also $(0,\bar{s})$-optimal. We note that in the given example $\mathbf{P}_x(\limsup_n g(x_n) = 0) = 1$ for all $x \in E$ and $\mathbf{P}_x(\tau^* < \infty) < 1$ for all $x < \gamma^*$.

Therefore the time τ^*, which is $(0,\bar{s})$-optimal, is not simultaneously $(0,s)$-optimal, since $\tau^* \notin \mathfrak{M}$.

§ 2. ε-optimal and optimal Markov times

1. Theorem 1 allows us to make a more detailed study of the problem of existence and structure of ε-optimal and optimal (in the three senses indicated above) Markov times in the case of functions $g \in L(A^-, A^+)$.

THEOREM 2. *Suppose the $\mathscr{B}$-measurable function $g(x)$ satisfies the conditions*

$$A^-: \mathbf{E}_x\left[\sup_n g^-(x_n)\right] < \infty, \qquad x \in E,$$
$$A^+: \mathbf{E}_x\left[\sup_n g^+(x_n)\right] < \infty, \qquad x \in E,$$

(*i.e.* $g \in L(A^-,A^+)$) *and* $v(x)$ *is its s.e.m.*

1) *For every* $\varepsilon > 0$ *the time*

$$\tau_\varepsilon = \inf\{n: v(x_n) \leqq g(x_n) + \varepsilon\}$$

is an ε-optimal stopping time (*more precisely, an (ε,s)-optimal stopping time*).

2) *The time*

$$\tau_0 = \inf\{n: v(x_n) = g(x_n)\}$$

is a $(0,\bar{s})$-optimal Markov time:

$$\tilde{\mathbf{E}}_x g(x_{\tau_0}) = \bar{s}(x) = v(x), \qquad x \in E.$$

3) *If the time* $\tau_0 = \inf\{n: v(x_n) = g(x_n)\}$ *is a stopping time* ($\tau_0 \in \mathfrak{M}$), *then it is optimal* (*more precisely*, $(0,s)$-*optimal*).

PROOF. 1) Under the assumption A^+

$$\limsup_n g(x_n) = \limsup_n v(x_n),$$

whence[8] $\mathbf{P}_x(\tau_\varepsilon < \infty) = 1$ for each $\varepsilon > 0$ and $x \in E$.

2) Applying Lemma 5 to the function $f(x) = v(x)$, we obtain that for each $N \in \boldsymbol{N}$

$$\begin{aligned} v(x) &= \int_{(\tau_0<N)} v(x_{\tau_0})\, d\mathbf{P}_x + \int_{(\tau_0\geqq N)} v(x_N)\, d\mathbf{P}_x \\ &= \int_{(\tau_0<N)} v(x_{\tau_0})\, d\mathbf{P}_x + \int_{(N\leqq\tau_0<N)} v(x_N)\, d\mathbf{P}_x + \int_{(\tau_0=\infty)} v(x_N)\, d\mathbf{P}_x \\ &\leqq \int_{(\tau_0<\infty)} \chi(\omega)_{(\tau_0<N)} \cdot v(x_{\tau_0})\, d\mathbf{P}_x + \int_{(N\leqq\tau_0<\infty)} \mathbf{E}_{x_N}\left[\sup_n g(x_n)\right] d\mathbf{P}_x \end{aligned}$$

---

8) Cf. the proof of Lemma 9.

$$+ \int_{(\tau_0=\infty)} \mathbf{E}_{x_N}\left[\sup_n g(x_n)\right] d\mathbf{P}_x \leqq \int_{(\tau_0<\infty)} \chi(\omega)_{(\tau_0<N)} \upsilon(x_{\tau_0}) d\mathbf{P}_x$$
$$+ \int_{(N\leqq\tau_0<\infty)} \sup_n g^+(x_n) d\mathbf{P}_x + \int_{(\tau_0=\infty)} \sup_{n\geqq N} g(x_n) d\mathbf{P}_x. \tag{2.31}$$

Hence, due to the condition $g \in L(A^-, A^+)$ and Fatou's Lemma,

$$\upsilon(x) \leqq \int_{(\tau_0<\infty)} \upsilon(x_{\tau_0}) d\mathbf{P}_x + \int_{(\tau_0=\infty)} \limsup_N g(x_N) d\mathbf{P}_x. \tag{2.32}$$

But

$$\int_{(\tau_0<\infty)} \upsilon(x_{\tau_0}) d\mathbf{P}_x = \int_{(\tau_0<\infty)} g(x_{\tau_0})\, d\mathbf{P}_x,$$

and by the condition A^+ and Lemma 6,

$$\int_{(\tau_0=\infty)} \limsup_n g(x_n) d\mathbf{P}_x = \int_{(\tau_0=\infty)} \limsup_n \upsilon(x_n) d\mathbf{P}_x.$$

Therefore from (2.32) we obtain

$$\upsilon(x) = \int_{(\tau_0<\infty)} g(x_{\tau_0}) d\mathbf{P}_x + \int_{(\tau_0=\infty)} \limsup_n g(x_n)\, d\mathbf{P}_x = \tilde{\mathbf{E}}_x g(x_{\tau_0}), \tag{2.33}$$

which proves the $(0,\bar{s})$-optimality of the time τ_0.

3) If $\tau_0 \in \mathfrak{M}$, then by the conditions A^- and A^+,

$$\int_{(\tau_0=\infty)} \limsup_n g(x_n) d\mathbf{P}_x = 0,$$

and from (2.33) we find

$$\upsilon(x) = \int_{(\tau_0<\infty)} g(x_{\tau_0})\, d\mathbf{P}_x = \int_\Omega g(x_{\tau_0})\, d\mathbf{P}_x = \mathbf{E}_x g(x_{\tau_0}) = s(x).$$

Consequently $\tau_0 \in \mathfrak{M}$, i.e. τ_0 is a $(0,s)$-optimal time.

REMARK 1. Let $\Gamma_\varepsilon = \{x\colon \upsilon(x) \leqq g(x) + \varepsilon\}$, $\varepsilon \geqq 0$. Then $\Gamma_\varepsilon \supseteq \Gamma_0$, $\varepsilon > 0$, where $\Gamma_\varepsilon \downarrow \Gamma_0$ as $\varepsilon \downarrow 0$.

REMARK 2. The condition A^+ involved in the statement of the theorem cannot, generally speaking, be weakened (see Example 5 below). However, the following result holds. Let $g \in L(A^-)$ and suppose that for given x_0

$$\mathbf{E}_{x_0}\left[\sup_n g^+(x_n)\right] < \infty.$$

Then the time τ_ε, $\varepsilon > 0$, is (ε,s)-optimal at the point x_0 $\left(\mathbf{E}_{x_0} g(x_{\tau_\varepsilon}) \geqq s(x_0) - \varepsilon\right)$, and the time τ_0 is $(0,\bar{s})$-optimal at the point x_0. Now if $\mathbf{P}_{x_0}(\tau_0 < \infty) = 1$, then τ_0 is an optimal stopping time (at the point x_0).

2. COROLLARY 1. *If the set E is finite, $-\infty < g(x) < \infty$, then there exists an optimal stopping time.*

Indeed, if E is a finite set, then, starting with some ε', we have $\Gamma_\varepsilon = \Gamma_0$ for $\varepsilon \leqq \varepsilon'$. Consequently the time $\tau_0 = \tau_\varepsilon$, $\varepsilon \leqq \varepsilon'$, is finite with $\mathbf{P}_x$-probability one, and, by part 3) of Theorem 2, $s(x) = \mathbf{E}_x g(x_{\tau_0})$, $x \in E$.

We consider several simple examples.

EXAMPLE 1. Let $E = \{0, 1, \ldots, N\}$ and suppose the transition probability $p(x,y) = \mathbf{P}_x(x_1 = y)$ satisfies

$$p(0,0) = p(N, N) = 1, \qquad p(i, i + 1) = p(i, i - 1) = 1/2, \quad i = 1, \ldots, N - 1.$$

The excessiveness of the function $f(x)$ in the given case implies its convexity,

$$f(x) \geqq \frac{f(x + 1) + f(x - 1)}{2}, \qquad x = 1, \ldots, N - 1,$$

where $f(0) = g(0)$ and $f(N) = g(N)$. Therefore the s.e.m. $v(x)$ for the function $g(x)$ is the smallest convex function "spread over" $g(x)$ with satisfaction of the end conditions $v(0) = g(0)$ and $v(N) = g(N)$.

Here an optimal stopping rule consists of breaking off observation at those points x for which $v(x) = g(x)$.

EXAMPLE 2. In contrast to the preceding example, where the states $\{0\}$ and $\{N\}$ were absorbing, we now assume that $p(0,1) = p(N, N - 1) = 1$. It can easily be seen that here the s.e.m. $v(x)$ is constant: $v(x) \equiv \max_x g(x)$, and an optimal stopping rule is to break off observation at the first time of hitting of points x for which $v(x) = g(x)$.

EXAMPLE 3. Again let $E = \{0, 1, \ldots, N\}$ and

$$p(0, 0) = 1, \quad p(N, N-1) = 1,$$
$$p(i, i + 1) = p(i, i - 1) = 1/2, \quad i = 1, \ldots, N - 1.$$

Then the s.e.m. $v(x)$ of the function g is the smallest "convex" hull of the function $g(x)$ satisfying the restrictions: $v(0) = g(0)$ and $v(x) \geqq g(x_0)$, $x \geqq x_0$, where x_0 is that (smallest) point where the function $g(x)$ achieves a maximum.

In the case of a finite number of states, as we have seen, an optimal stopping time exists. But if the set of states is countable, this is generally not so. We illustrate this situation in the following simple example.

EXAMPLE 4. Let $E = \{0, 1, \ldots\}$ and $p(i, i+1) = 1$ (i.e., we are considering deterministic motion to the right) and let $g(x) \geqq 0$ be a monotone increasing function such that $K = \lim_{x\to\infty} g(x) < \infty$. Since here

$$\varphi(x) = \mathbf{E}_x \left[\sup_n g(x_n)\right] \equiv K,$$

we have $G^N\varphi(x) \equiv K$, and consequently, by Lemma 8, the s.e.m. $v(x) \equiv K$ (which, however, was *a priori* obvious). It is now easy to see that an optimal stopping time does not exist, while the time $\tau \equiv \infty$ is $(0,\tilde{s})$-optimal since

$$\mathbf{P}_x\left(\limsup_n g(x_n) = K\right) = 1 \qquad \text{for all } x \in E.$$

On the other hand, it is clear that the time $\tau_\varepsilon = \inf\{n: x_n \geqq K - \varepsilon\}$ is an ε-optimal stopping time for each $\varepsilon > 0$.

3. We might think that in the case where the condition A^+ is violated the times

$$\tau_\varepsilon = \min\{n: v(x_n) \leqq g(x_n) + \varepsilon\}$$

still remain ε-optimal. However, generally speaking, this is not so, as can be seen from the following example.

EXAMPLE 5. Let $E = \{0,1,2,\dots\}$,

$$p(0, 0) = 1, \quad p(i, i + 1) = p(i, i - 1) = 1/2,$$
$$i = 1, 2, \dots,$$

and $g(0) = 1$, $g(i) = i$, $i = 1,2,\dots$. It can be shown that here $\mathbf{E}_x[\sup_n g(x_n)] = \infty$, $x = 1,2,\dots$. Using Lemma 4, it is easy to find the s.e.m. $v(x)$ of the function $g(x)$: $v(0) = 1$, $v(x) = x + 1$, $x = 1,2,\dots$. For $0 \leqq \varepsilon < 1$ the set $\Gamma_\varepsilon = \{x: v(x) \leqq g(x) + \varepsilon\}$ consists of the single point $\{0\}$ and the time $\tau_\varepsilon = \inf\{n: x_n = 0\}$ is finite with probability one for all $x \in E$. Therefore $\mathbf{E}_x g(x_{\tau_\varepsilon}) = 1$, $x \in E$. But, on the other hand, it is clear that the time $\tilde{\tau} = \inf\{n: x_n \in E\}$, which prescribes immediate stopping, yields a "reward" at any point $x = 2,3,\dots$ of $\mathbf{E}_x g(x_{\tilde{\tau}}) = g(x)$, equal respectively to $2,3,\dots$, i.e., larger than the "reward" from stopping prescribed by the time τ_ε.

4. In the case where the condition A^+ is violated we cannot, generally speaking, find a time $\sigma_\varepsilon \in \mathfrak{M}$ such that for all points x where $s(x) < \infty$

$$s(x) \leqq \mathbf{E}_x g(x_{\sigma_\varepsilon}) + \varepsilon.$$

Nevertheless we have the following result.

COROLLARY 2. *For each fixed value x_0, where $s(x_0) < \infty$, and each $\varepsilon > 0$ there exists (by the definition of* sup*) a time $\sigma_\varepsilon(x_0)$ such that*

$$s(x_0) \leqq \mathbf{E}_{x_0} g(x_{\sigma_\varepsilon(x_0)}) + \varepsilon.$$

This time can be constructed as follows. Let

$$g^b(x) = \min(b, g(x)), \qquad s^b(x) = \sup_{\tau \in \mathfrak{M}} \mathbf{E}_x g^b(x_\tau).$$

We already know that $s^b(x) \uparrow s(x)$. Thus for the point x_0 where $s(x_0) < \infty$, and $\varepsilon > 0$, we can find an $N = N(x_0, \varepsilon)$ such that for all $n \geqq N$

$$s(x_0) - s^b(x_0) \leqq \varepsilon/2. \tag{2.34}$$

It follows from the boundedness of the function $g^b(x)$ that the s.t.

$$\sigma_\varepsilon(x_0) = \inf \{n: g^{N(x_0,\varepsilon)}(x_n) + \varepsilon/2 \geqq s^{N(x_0,\varepsilon)}(x_n)\}$$

is such that $\sigma_\varepsilon(x_0) \in \mathfrak{M}$ and

$$\mathbf{E}_x g^N (x_{\sigma_\varepsilon(x_0)}) \geqq s^N (x) - \varepsilon/2, \qquad x \in E. \tag{2.35}$$

We obtain the desired result from (2.34) and (2.35), since

$$s(x_0) \geqq \mathbf{E}_{x_0} g(x_{\xi_\varepsilon(x_0)}) \geqq \mathbf{E}_{x_0} g^N (x_{\sigma_\varepsilon(x_0)}) \geqq s^N (x_0) - \varepsilon/2 \geqq s(x_0) - \varepsilon.$$

5. Up to now, in both § 1 and § 2 we assumed that the Markov chain $X = (x_n, \mathscr{F}_n, \mathbf{P}_x)$, $n \in \boldsymbol{N}$, is nonterminating.

The case of terminating Markov chains $X = (x_n, \zeta, \mathscr{F}_n, \mathbf{P}_x)$, $n \in \boldsymbol{N}$, does not require the introduction of new ideas and can be investigated in an analogous fashion. We deal with only those changes which have to be made in the basic statements.

By the class $\overline{\mathfrak{M}}$ we mean Markov times τ such that $\mathbf{P}_x(\tau \leqq \zeta) = 1$ for all $x \in E$. The class $\mathfrak{M}$ is defined as the collection of those times τ for which $\mathbf{P}_x(\tau < \zeta) = 1$, $x \in E$. In accord with § 2 of Chapter I, we let Δ denote a fictitious state and extend the definition of the function $f \in L(A^-)$ to the set $E_\Delta = E \cup \{\Delta\}$ with the aid of the condition $f(\Delta) = 0$. For $g \in L(A^-)$ we let

$$\mathbf{E}_x g(x_\tau) = \mathbf{E}_x[g(x_\tau); \tau < \zeta] = \int_{(\tau<\zeta)} g(x_\tau) d\,\mathbf{P}_x,$$

$$\tilde{\mathbf{E}}_x g(x_\tau) = \mathbf{E}_x [g(x_\tau); \tau < \zeta] + \mathbf{E}_x \left[\limsup_n g(x_n); \tau = \infty\right].$$

From these definitions and the condition $g(\Delta) = 0$ it is obvious that

$$\mathbf{E}_x g(x_\tau) = \int_{(\tau<\infty)} g(x_\tau) d\mathbf{P}_x,$$

$$\tilde{\mathbf{E}}_x g(x_\tau) = \int_{(\tau<\infty)} g(x_\tau) d\mathbf{P}_x + \int_{(\tau=\infty)} \limsup_n g(x_n) d\,\mathbf{P}_x, \tag{2.36}$$

and therefore the "exterior" cases of nonterminating and terminating Markov chains are not different. By analogy with (2.2) and (2.3) we let

$$s(x) = \sup_{\tau \in \mathfrak{M}} \mathbf{E}_x g(x_\tau), \qquad \bar{s}(x) = \sup_{\tau \in \overline{\mathfrak{M}}} \mathbf{E}_x g(x_\tau),$$

$$\tilde{s}(x) = \sup_{\tau \in \overline{\mathfrak{M}}} \tilde{\mathbf{E}}_x g(x_\tau), \tag{2.37}$$

where $\tilde{s}(x)$ is introduced only for nonnegative functions $g(x)$.

Looking over the proof of Lemmas 1—8, it is easy to see that they remain valid for terminating Markov chains. Theorems 1 and 2 remain valid analogously.

6. In the conclusion of this section we give a solution of the problem, noted in the introduction, of choosing the best object. We recall the conditions of this problem.

Suppose we have n objects, indexed by the numbers $1,2,\dots,n$ so that, say, the object numbered 1 is classified as the "best," and the one numbered n as the "worst."

We assume that the objects are examined in random order, and as a result of comparison of any two objects it becomes clear which is better, but their true indices remain unknown. The problem consists of choosing, with maximum probability, the "best object," i.e., the object with index 1. The essence of this problem is that an object which is examined is either rejected (and then we cannot return to it) or accepted (and then the process of choosing is terminated). We also assume that if an object is rejected, then we also reject all succeeding ones worse than this one[9] (not taking into account the object in the last position, which must be accepted if a choice was not made earlier).

Let $(a_1,\dots,a_n)$ be any of the $n!$ permutations of the numbers $(1,2,\dots,n)$. The assumption of random order of the collection of objects means that the probability of any permutation $(a_1,\dots,a_n)$ is $1/n!$.

Let σ denote the time of choice of the (alleged "best") object. The problem is how to find a time σ^* for which

$$s(1) = \mathbf{P}(a_{\sigma^*} = 1) = \sup_{\sigma} \mathbf{P}(a_\sigma = 1)$$

(from the obvious condition $\sigma \leqq n$ it is easy to see that an optimal time σ^* indeed exists; besides, this will also follow from the discussion below).

With each process of inspecting objects we associate a sequence $x_0, x_1, \dots,$ where $x_0 = 1$, and x_{i+1} is the order index of an object better than the object with order index x_i, $i \geqq 0$. Thus, for example, x_1 is the index of the place in which we have the first object better than the object which is in the first place among those being examined. Here, if an object better than the object with order index x_i does not exist, then x_{i+1} is not defined. It is obvious that $1 \leqq x_i \leqq n$ and $x_{i+1} \geqq x_i$.

Relatively simple combinatorial computations show[10] that for $1 \le b_1 < \dots < b_i < b_{i+1}$

$$\begin{aligned}\mathbf{P}(x_{i+1} = b_{i+1} \mid x_1 = b_1, \dots, x_i = b_i) \\ = \mathbf{P}(x_{i+1} = b_{i+1} \mid x_i = b_i) \frac{b_i}{b_{i+1}\,(b_{i+1}-1)}.\end{aligned}$$

Hence it follows that the sequence $x_0 = 1, x_1, \dots$ forms a homogeneous (terminating) Markov chain in the phase space $E = \{1,2,\dots,n\}$ with transition probabilities

[9] The consideration of only such procedures for choosing the best object does not restrict generality, as shown in [27], Chapter III, § 1.

[10] For more detail see [27], loc. cit.

$$p(x, y) = \begin{cases} \dfrac{x}{y(y-1)}, & x < y, \\ 0, & x \geqq y. \end{cases} \tag{2.38}$$

We note that $\sum_{y>x} p(x,y) = 1 - x/n$. Therefore with probability x/n the given Markov chain terminates in the state x, which happens only when the object with order index x is best. In other words, $\mathbf{P}_x\{\zeta = 1\} = x/n$.

We let $P_1(t)$ denote the probability that the object with order index x_t is best ($x_0 = 1$, $t = 1,\dots,n$). Obviously

$$P_1(t) = \sum_{y>1} \frac{y}{n} \mathbf{P}(x_t = y | x_0 = 1) = \mathbf{E}_1 \frac{x_t}{n}.$$

Thus the probability of interest is $s(1) = \sup \mathbf{E}_1(x_\tau/n)$, where the upper bound is taken over all stopping times $\tau \leqq n$.

We let

$$s(x) = \sup \mathbf{E}_x(x_\tau/n), \qquad \Gamma_0 = \{x: s(x) = g(x)\},$$

where $g(x) = x/n$. By Theorem 2 the time $\tau^* = \min\{t \leqq n: x_t \in \Gamma_0\}$ is an optimal s.t. To find it, we turn to the recursion equations

$$s(x) = \max\{g(x), Ts(x)\}. \tag{2.39}$$

In this problem

$$s(x) = \max\left\{\frac{x}{n}, \sum_{y=x}^{n-1} \frac{x}{y} \frac{s(y+1)}{y+1}\right\}, \qquad x = 1,\dots, n-1, \tag{2.40}$$

where, obviously, $s(n) = 1$. Making use of this relationship, it is easy to find the values of the function $s(x)$ for all $x \in E$, successively finding $s(n), s(n-1),\dots,s(x)$. We have

$$s(n-1) = \max\left\{\frac{n-1}{n}, \frac{1}{n}\right\} = \frac{n-1}{n} = g(n-1),$$

$$s(n-2) = \max\left\{\frac{n-2}{n}, \frac{n-2}{n}\left(\frac{1}{n-1} + \frac{1}{n-2}\right)\right\} = \frac{n-2}{n} = g(n-2),$$

$$\dots\dots\dots\dots\dots\dots\dots\dots\dots\dots\dots\dots\dots$$

$$s(x) = \max\left\{\frac{x}{n}, \frac{x}{n}\left(\frac{1}{n-1} + \dots + \frac{1}{x}\right)\right\} = \frac{x}{n} = g(x),$$

if

$$1/(n-1) + \dots + 1/x \leqq 1.$$

Suppose $k^* = k^*(n)$ is determined from the inequalities

$$\frac{1}{n-1}+\frac{1}{n-2}+\ldots+\frac{1}{k^*}\leqq 1<\frac{1}{n-1}+\frac{1}{n-2}+\ldots+\frac{1}{k^*-1}$$

(we note that $k^*(n) \sim n/e$ for large n). Then $s(x) = g(x)$ for all $x \geqq k^*$ and $s(x) > g(x)$ for $x < k^*$. Consequently $\Gamma_0 = \{k^*, k^* + 1,\ldots,n\}$, whence it follows that an optimal rule for choosing the best object is to examine and pass over $k^* - 1$ objects and then choose the first object (at the time σ^*) which is better than all the preceding ones.

We shall find the probability $s(1) = \mathbf{P}(a_{\sigma^*} = 1)$ of choosing the best object following this optimal rule. For all $x < k^*$

$$s(x) = x\sum_{y=x}^{n-1}\frac{s(y+1)}{y(y+1)},$$

where $s(y) = y/n$, $y \geqq k^*$. Hence it is easy to find that

$$s(1) = \ldots = s(k^*-1) = \frac{k^*-1}{n}\left\{\frac{1}{k^*-1}+\ldots+\frac{1}{n-1}\right\}.$$

Since $k^*(n) \sim n/e$ for large n, then $s(1) \sim 1/e$, where $1/e \approx 0.368$.

Thus for sufficiently large n we can choose the best object with probability approximately equal to 0.368, even though it might seem at first glance that as the number of objects being inspected increases the probability $s(1)$ should approach zero.

§ 3. Excessive characterization of value. General case

1. As follows from Theorem 1, for functions $g \in L(A^-)$ the value $s(x)$ is the smallest excessive majorant of the function $g(x)$. Does Theorem 1 hold if the condition A^- is violated? We give an example which shows that in the general case the value $s(x)$ may not be the smallest excessive majorant of $g(x)$.

EXAMPLE. Suppose the phase space $E = \{0,2,2^2,\ldots\}$ and the Markov chain $X = (x_n,\mathscr{F}_n,\mathbf{P}_x)$, $n \in \boldsymbol{N}$, has a structure such that

$$x_{n+1} = 2x_n\cdot\xi_{n+1},$$

where $\{\xi_n\}$, $n \in \boldsymbol{N}$, is a sequence of independent identically distributed random variables, $\mathbf{P}(\xi_n = 0) = \mathbf{P}(\xi_n = 1) = 1/2$, $\mathscr{F}_n = \sigma\{\omega\colon x_0, x_1, \ldots, x_n\}$, and the measure $\mathbf{P}_x$ is defined in the natural fashion. In other words, a "particle" leaving from the point $x \in E$ hits with equal probability either the point $2x$ or the point 0, where it remains.

We let $g(x) = -x$ and consider the value

$$s(x) = \sup_{\tau\in\mathfrak{M}}\mathbf{E}_x g(x_\tau).$$

Since $g(x) \leqq 0$, the mathematical expectation $\mathbf{E}_x g(x_\tau)$ is defined for each $\tau \in \mathfrak{M}$. It is easy to see that for the given example

$$\mathbf{E}_x\Big[\sup_n g^-(x_n)\Big] = \mathbf{E}_x\Big[\sup_n x_n\Big] = \infty, \qquad x \neq 0,$$

and thus the condition A^- is violated.

To investigate the function $g \in L$ we find its s.e.m. $v(x)$. It can easily be verified that $Tg(x) = g(x)$, $x \in E$. Consequently the function $v(x) = g(x)$ is the smallest excessive majorant of $g(x)$, and if Theorem 1 were to remain valid, we would have $s(x) = g(x) = -x$. However, $s(x) \equiv 0$. Indeed, let $\tau = \inf\{n\colon x_n = 0\}$. Since $\mathbf{P}_x(\bar{\tau} < \infty) = 1$, $x \in E$, we have $\bar{\tau} \in \mathfrak{M}$. But $\mathbf{E}_x g(x_{\bar{\tau}}) = 0$ and, obviously, $s(x) \leqq 0$. Therefore the time $\bar{\tau}$ is optimal, $s(x) \equiv 0$ and $s(x) \neq v(x)$, where $v(x)$ is the s.e.m.

It is easy to explain the fact that here $s(x) \neq v(x)$. The point is that in the proof of Theorem 1 we essentially worked on the basis of Lemma 1, which, as the above example shows, may not be true when the condition A^- is violated. It is therefore natural to think that in the general case an "excessive" characterization of the value should be sought in the class of those e.f. for which Lemma 1 holds.

With this aim, we introduce some notation and definitions.

Suppose the function $g \in L$. Let $\mathfrak{N}_g \subseteq \mathfrak{M}$ denote the class of stopping times $\tau \in \mathfrak{M}$ for which the mathematical expectation $\mathbf{E}_x g(x_\tau)$ is defined for all $x \in E$ (i.e. for each $x \in E$, either $\mathbf{E}_x g^-(x_\tau) < \infty$ or $\mathbf{E}_x g^+(x_\tau) < \infty$). Analogously, let $\bar{\mathfrak{N}}_g \subseteq \bar{\mathfrak{M}}$ be the class of Markov times $\tau \in \bar{\mathfrak{M}}$ for which the mathematical expectation $\bar{\mathbf{E}}_x g(x_\tau)$ is defined for all $x \in E$.

It is clear that $\mathfrak{N}_g = \mathfrak{M}$ and $\bar{\mathfrak{N}}_g = \bar{\mathfrak{M}}$ if $g \in L\,(A^-)$. We further let

$$s(x) = \sup_{\tau \in \mathfrak{N}_g} \mathbf{E}_x g(x_\tau), \tag{2.41}$$

$$\bar{s}(x) = \sup_{\tau \in \bar{\mathfrak{N}}_g} \bar{\mathbf{E}}_x g(x_\tau). \tag{2.42}$$

We call each of the functions $s(x)$ and $\bar{s}(x)$ a *value*, and, analogous to the case $g \in L\,(A^-)$, we define the (ε,s)- and $(\varepsilon,\bar{s})$-optimal Markov times.

We let $\mathfrak{N}_g$ $(\bar{\mathfrak{N}}_g)$ denote the class of those Markov times for which $\mathbf{E}_x g^-(x_\tau) < \infty$ $(\bar{\mathbf{E}}_x g^-(x_\tau) < \infty)$ for all $x \in E$. It is important to note that in (2.41) and (2.42) it is sufficient to take the upper bound not over all $\tau \in \mathfrak{N}_g$ and $\tau \in \bar{\mathfrak{N}}_g$, but only over $\tau \in \mathfrak{N}_g$ and $\tau \in \bar{\mathfrak{N}}_g$ respectively. Indeed, if $\mathbf{E}_{x_0} g^-(x_\tau) = +\infty$ at the point $x_0 \in E$, then we set $\bar{\tau}(\omega) \equiv 0$. If $\mathbf{E}_{x_0} g^-(x_\tau) < \infty$, then we define

$$\bar{\tau}(\omega) = \begin{cases} \tau(\omega) & \text{if } x_0(\omega) = x_0, \\ 0 & \text{if } x_0(\omega) \neq x_0. \end{cases}$$

Since $\{x_0\} \in \mathscr{B}$, it follows that $\bar{\tau}(\omega)$ is a Markov time. Moreover $\bar{\tau} \in \mathfrak{N}_g$ and $\mathbf{E}_{x_0} g(x_\tau) \leqq \mathbf{E}_{x_0} g(x_{\bar{\tau}})$. Hence for all $x \in E$

$$\sup_{\tau \in \mathfrak{N}_g} \mathbf{E}_x g(x_\tau) = \sup_{\tau \in \mathfrak{N}_g} \mathbf{E}_x g(x_\tau).$$

Analogously we can prove the equality

$$\sup_{\tau \in \bar{\mathfrak{N}}_g} \tilde{\mathbf{E}}_x g(x_\tau) = \sup_{\tau \in \bar{\mathfrak{M}}_g} \tilde{\mathbf{E}}_x g(x_\tau).$$

DEFINITION 1. The excessive function $f \in \mathscr{E}$ is said to be *$\bar{\mathfrak{N}}_g$-regular* if $\tilde{\mathbf{E}}_x f(x_\tau)$ is defined for all $\tau \in \mathfrak{N}_g$ and

$$f(x) \geqq \tilde{\mathbf{E}}_x f(x_\tau). \tag{2.43}$$

DEFINITION 2. The excessive function $f \in \mathscr{E}$ is said to be *$\bar{\mathfrak{M}}_g$-regular* if $\tilde{\mathbf{E}}_x f^-(x_\tau) < \infty$ for all $\tau \in \bar{\mathfrak{M}}_g$ and (2.43) is satisfied.

Every $\bar{\mathfrak{N}}_g$-regular excessive majorant of the function $g(x)$ is at the same time $\bar{\mathfrak{M}}_g$-regular, and therefore the class of $\bar{\mathfrak{M}}_g$-regular e.m.'s of the function $g(x)$ is wider than the class of $\bar{\mathfrak{N}}_g$-regular e.m.'s.

For brevity, $\bar{\mathfrak{M}}_g$-regular functions will also be called simply *regular*.

We shall omit the subscript g in all those cases where we consider a completely determined function $g \in L$ for $\mathfrak{N}_g$, $\mathfrak{M}_g$, $\bar{\mathfrak{N}}_g$ and $\bar{\mathfrak{M}}_g$.

2. The basic result on excessive characterization of the values $s(x)$ and $\tilde{s}(x)$ defined in (2.41) and (2.42) is contained in the following theorem.

THEOREM 3. *Suppose the function $g \in L$. Then the value $s(x)$ is the smallest regular excessive majorant of the function $g(x)$, $s(x) = \tilde{s}(x)$ and*[11] $s(x) = \max\{g(x), Ts(x)\}$.

For functions $g \in L(A^-)$ the assertion of the theorem follows from Theorem 1 and Lemma 1. Now we shall establish it for functions $g \in L(A^+)$.

LEMMA 10. *Suppose the function $g \in L(A^+)$. Then $s(x)$ is the smallest $\bar{\mathfrak{N}}$-regular excessive majorant of the function $g(x)$, $s(x) = \tilde{s}(x)$ and*

$$s(x) = \lim_{N \to \infty} G^N \varphi(x), \qquad \varphi(x) = \mathbf{E}_x \left[\sup_n g(x_n)\right].$$

PROOF. Let

$$g_a(x) = \max\big(a, g(x)\big), \quad a \leqq 0, \quad s_a(x) = \sup_{\tau \in \mathfrak{N}} \mathbf{E}_x g_a(x_\tau),$$
$$\tilde{s}_a(x) = \sup_{\tau \in \bar{\mathfrak{M}}} \tilde{\mathbf{E}}_x g_a(x_\tau), \quad \bar{s}_a(x) = \sup_{\tau \in \bar{\mathfrak{M}}} \mathbf{E}_x g_a(x_\tau).$$

By Theorem 1, $s_a(x) = \tilde{s}_a(x)$, where (see Lemma 1) $s_a(x)$ is the smallest regular excessive majorant of the function $g_a(x)$.

Since $s_a(x)$ does not increase as $a \to -\infty$, the limit $s_*(x) = \lim_{a \to -\infty} s_a(x)$ exists, and obviously

$$s_*(x) \geqq \tilde{s}(x) \geqq s(x). \tag{2.44}$$

According to Theorem 1

---

[11] If $g(x) \geqq 0$, then from the equality $s(x) = \tilde{s}(x)$ it obviously also follows that $s(x) = \tilde{s}(x) = \bar{s}(x)$.

$$s_a(x) = \max\{g_a(x), Ts_a(x)\}. \tag{2.45}$$

Since $s_a(x) \downarrow s_*(x)$ as $a \downarrow -\infty$ and

$$\mathbf{E}_x\left[\sup_n g^+(x_n)\right] < \infty,$$

from (2.45) as $a \downarrow -\infty$ we obtain the equation

$$s_*(x) = \max\{g(x), Ts_*(x)\}. \tag{2.46}$$

Since, moreover, for $\tau \in \bar{\mathfrak{N}}$

$$s_a(x) \geqq \tilde{\mathbf{E}}_x s_a(x_\tau) \geqq \tilde{\mathbf{E}}_x s_*(x_\tau),$$

we have

$$s_*(x) \geqq \tilde{\mathbf{E}}_x s_*(x_\tau). \tag{2.47}$$

Consequently the constructed function $s_*(x) \geqq \bar{s}(x)$, satisfies (2.46) and is an $\bar{\mathfrak{N}}$-regular excessive majorant of $g(x)$.

Now we deal with the inequality $s_*(x) \leqq s(x)$ which, together with (2.44), leads to the chain of equalities $s_*(x) = \bar{s}(x) = s(x)$ and $s_*(x) = \bar{s}(x) = \bar{s}(x) = s(x)$, if $g \geqq 0$.

By Lemma 6,

$$\begin{aligned}\limsup_n s_a(x_n) &= \limsup_n g_a(x_n)\\ &= \max\left(\limsup_n g(x_n), a\right), \qquad (\mathbf{P}_x\text{-a.s.}, x \in E),\end{aligned}$$

where, as $a \downarrow -\infty$,

$$\max\left(\limsup_n g(x_n), a\right) \downarrow \limsup_n g(x_n),$$
$$\limsup_n s_n(x_n) \geqq \limsup_n s_*(x_n).$$

Consequently

$$\limsup_n g(x_n) \geqq \limsup_n s_*(x_n), \qquad (\mathbf{P}_x\text{-a. s., } x \in E).$$

But $g(x) \leqq s_*(x)$ according to (2.44), and therefore ($\mathbf{P}_x$-a.s., $x \in E$)

$$\limsup_n g(x_n) = \limsup_n s_*(x_n). \tag{2.48}$$

For $\varepsilon > 0$ we let

$$\tau_\varepsilon = \inf\{n \colon s_*(x_n) \leqq g(x_n) + \varepsilon\}$$

and we show that $\mathbf{P}_x(\tau_\varepsilon < \infty) = 1$, $x \in E$, i.e. $\tau_\varepsilon \in \mathfrak{M}$. To this end we apply Lemma 5 to $s_*(x)$. Then for $N \in \boldsymbol{N}$ and $0 \geqq a > -\infty$

$$\begin{aligned}
-\infty < s_*(x) &= \int_{(\tau_\varepsilon<N)} s_*(x_{\tau_\varepsilon})d\,\mathbf{P}_x + \int_{(\tau_\varepsilon\geqq N)} s_*(x_N)d\,\mathbf{P}_x \\
&= \int_{(\tau_\varepsilon<\infty)} \chi_{(\tau_\varepsilon<N)} s_*(x_{\tau_\varepsilon})d\,\mathbf{P}_x + \int_\Omega s_*(x_N)\chi_{(\tau_\varepsilon\geqq N)}d\mathbf{P}_x \\
&\leqq \int_{(\tau_\varepsilon<\infty)} \chi_{(\tau_\varepsilon<\infty)} s_*(x_{\tau_\varepsilon})d\,\mathbf{P}_x + \int_\Omega s_a(x_N)\chi_{(\tau_\varepsilon\geqq N)}d\,\mathbf{P}_x \\
&\leqq \int_{(\tau_\varepsilon<\infty)} \chi_{(\tau_\varepsilon<N)} s_*(x_{\tau_\varepsilon})d\,\mathbf{P}_x + \int_\Omega \mathbf{E}_{x_N}\Big[\sup_n g_a(x_n)\Big]\chi_{(\tau_\varepsilon\geqq N)}d\mathbf{P}_x \\
&\leqq \int_{(\tau_\varepsilon<\infty)} \chi_{(\tau_\varepsilon<N)} s_*(x_{\tau_\varepsilon})d\,\mathbf{P}_x + \int_\Omega \sup_{n\geqq N} g_a(x_n)\chi_{(\tau_\varepsilon\geqq N)}d\mathbf{P}_x.
\end{aligned}$$

Since $s_*(x_{\tau_\varepsilon}) \leqq \mathbf{E}_{x_{\tau_\varepsilon}}[\sup_n g^+(x_n)]$, $\chi_{(\tau_\varepsilon<N)}(\omega) \to \chi_{(\tau_\varepsilon<\infty)}(\omega)$, as $N\to\infty$, and $g \in L(A^+)$, by Fatou's Lemma we have

$$-\infty < s_*(x) \leqq \int_{(\tau_\varepsilon<\infty)} s_*(x_{\tau_\varepsilon})\,d\mathbf{P}_x + \int_{(\tau_\varepsilon=\infty)} \limsup_n g_a(x_n)d\mathbf{P}_x.$$

As $a \to -\infty$

$$\limsup_n g_a(x_n) \downarrow \limsup_n g(x_n) \qquad (\mathbf{P}_x\text{-a. s., } x\in E),$$

where

$$\limsup_n g_a(x_n) \leqq \sup g^+(x_n) \qquad (\mathbf{P}_x\text{-a.s., } x\in E).$$

Therefore

$$\lim_{a\to-\infty}\int_{(\tau_\varepsilon=\infty)} \limsup_n g_a(x_n)d\mathbf{P}_x = \int_{(\tau_\varepsilon=\infty)} \limsup_n g(x_n)d\mathbf{P}_x,$$

and consequently

$$-\infty < s_*(x) \leqq \int_{(\tau_\varepsilon<\infty)} s_*(x_{\tau_\varepsilon})d\,\mathbf{P}_x + \int_{(\tau_\varepsilon=\infty)} \limsup_n g(x_n)d\mathbf{P}_x.$$

From this inequality it follows that on the set $\{\omega\colon \tau_\varepsilon = \infty\}$

$$\limsup_n g\big(x_n(\omega)\big) > -\infty \qquad (\mathbf{P}_x\text{-a.s., } x\in E).$$

In view of the condition $g \in L(A^+)$ it is also clear that

$$\limsup_n g\big(x_n(\omega)\big) < \infty \qquad (\mathbf{P}_x\text{-a. s., } x\in E).$$

On the set $\{\omega\colon \tau_\varepsilon(\omega) = \infty\}$

$$\limsup_n s_*\big(x_n(\omega)\big) > \limsup_n g\big(x_n(\omega)\big).$$

Therefore

$$
\begin{aligned}
&\mathbf{P}_x(\tau_\varepsilon(\omega) = \infty) \\
&\quad = \mathbf{P}_x\left\{\left[\limsup_n s_*(x_n) > \limsup_n g(x_n)\right] \cap [\tau_\varepsilon = \infty]\right\} \\
&\quad = \mathbf{P}_x\left\{\left[\limsup_n s_*(x_n) > \limsup_n g(x_n)\right] \cap [\tau_\varepsilon = \infty]\right. \\
&\qquad\qquad \left.\cap \left[\left|\limsup_n g(x_n)\right| \neq \infty\right]\right\} \\
&\quad \leqq \mathbf{P}_x\left\{\left[\limsup_n s_*(x_n) > \limsup_n g(x_n)\right] \cap \left[\left|\limsup_n g(x_n)\right| \neq \infty\right]\right\}.
\end{aligned}
$$

But by (2.48)

$$
\mathbf{P}_x\left\{\left[\limsup_n s_*(x_n) > \limsup_n g(x_n)\right] \cap \left[\left|\limsup_n g(x_n)\right| \neq \infty\right]\right\} = 0
$$

and consequently $\mathbf{P}_x(\tau_\varepsilon = \infty) = 0$, $x \in E$, i. e. $\tau_\varepsilon \in \mathfrak{M}$.

Due to the condition A^+ we have $\tau_\varepsilon \in \mathfrak{N}$. Therefore

$$
s(x) \geqq \mathbf{E}_x g(x_{\tau_\varepsilon}) \geqq \mathbf{E}_x s_*(x_{\tau_\varepsilon}) - \varepsilon. \tag{2.49}
$$

We shall show that

$$
\mathbf{E}_x s_*(x_{\tau_\varepsilon}) \geqq s_*(x). \tag{2.50}
$$

Since $s_*(x) \geqq g(x)$, $s_*^-(x) \leqq g^-(x)$. But by the definition of the class L it follows that $\mathbf{E}_x g^-(x_n) < \infty$ for all $x \in E$ and $n \in \boldsymbol{N}$, and consequently $\mathbf{E}_x s_*^-(x_n) < \infty$, $n \in \boldsymbol{N}$, $x \in E$. On the other hand, it follows from condition A^+ that $\mathbf{E}_x s_*^+(x_n) < \infty$, $n \in \boldsymbol{N}$, $x \in E$. Therefore the function $s_*(x)$, which satisfies the equation (2.46), also has the property that $\mathbf{E}_x \left| s_*(x_n) \right| < \infty$ for all $n \in \boldsymbol{N}$ and $x \in E$. We again apply Lemma 5 to $s_*(x)$. Then for $n \in \boldsymbol{N}$ we obtain that for all $a \leqq 0$

$$
\begin{aligned}
&\int_{\tau_\varepsilon < N} s_*(x_{\tau_\varepsilon})\, d\mathbf{P}_x \\
&\quad = s_*(x) - \int_{(\tau_\varepsilon \geqq N)} s_*(x_N) d\mathbf{P}_x \geqq s_*(x) - \int_{(\tau_\varepsilon \geqq N)} s_a(x_N) d\mathbf{P}_x \\
&\quad \geqq s_*(x) - \int_{(\tau_\varepsilon \geqq N)} s_a^+(x_N) d\mathbf{P}_x \geqq s_*(x) - \int_{(\tau_\varepsilon \geqq N)} \sup_n g^+(x_n) d\mathbf{P}_x.
\end{aligned} \tag{2.51}
$$

But

$$
\mathbf{P}_x(\tau_\varepsilon < \infty) = 1, \qquad \mathbf{E}_x\left[\sup_n g^+(x_n)\right] < \infty, \qquad x \in E.
$$

Therefore, passing to the limit in (2.51) as $N \to \infty$, we obtain (2.50), which together with (2.49) yields the inequality $s(x) \geqq s_*(x) - \varepsilon$. Since $\varepsilon > 0$ was arbitrary, we have $s(x) \geqq s_*(x)$ and $s(x) = s_*(x)$.

To complete the proof of the lemma we still have to show that $s(x)$ is the smallest $\mathfrak{N}$-regular e.m. and that $s(x) = \lim_{N \to \infty} G^N \varphi(x)$.

Suppose $f(x)$ is an $\mathfrak{R}$-regular e.m. of the function $g(x)$. Then $\mathbf{E}_x f(x_\tau)$ is defined and

$$f(x) \geqq \mathbf{E}_x f(x_\tau) \geqq \mathbf{E}_x g(x_\tau),$$

and consequently $f(x) \geqq s(x)$. Hence it follows that $s(x)$ is the smallest $\mathfrak{R}$-regular e.m. of $g(x)$.

Let

$$\bar{v}(x) = \lim_{N\to\infty} G^N\varphi(x), \qquad \varphi_a(x) = \mathbf{E}_x\left[\sup_n g_a(x_n)\right], \quad a \leqq 0,$$
$$G_a\varphi_a(x) = \max\{g_a(x), T\varphi_a(x)\}.$$

Then

$$G_a^N\varphi_a(x) \geqq G^N\varphi(x) \geqq \bar{v}(x).$$

By Lemma 9, $s_a(x) = \lim_{N\to\infty} G_a^N\varphi_a(x)$ and consequently $s_*(x) = \lim_{a\to-\infty} s_a(x) \geqq \bar{v}(x)$. To prove the reverse inequality we note that

$$Gs_*(x) = \max\{g(x), Ts_*(x)\} = s_*(x)$$

and consequently $G^N s_*(x) = s_*(x)$. Further, obviously $s_*(x) \leqq \varphi(x)$. Therefore

$$Gs_*(x) \leqq G\varphi(x), \qquad s_*(x) = G^N s_*(x) \leqq G^N\varphi(x),$$

whence $s_*(x) \leqq \bar{v}(x)$. The lemma has been proved.

REMARK 1. Suppose $g \in L(A^+)$ and $g_a(x) = \max(a, g(x))$, where $a \leqq 0$. Then the smallest $\mathfrak{R}$-regular excessive majorant of the function $g(x)$ is

$$s(x) = \lim_{a\to-\infty}\lim_{N\to\infty} G_a^N\varphi_a(x), \tag{2.52}$$

where

$$\varphi_a(x) = \mathbf{E}_x\left[\sup_n g_a(x_n)\right], \qquad G_a f(x) = \max\{g_a(x), Tf(x)\}.$$

REMARK 2. Suppose $g \in L(A^+)$ and

$$g_a^b(x) = \begin{cases} b, & g(x) > b, \\ g(x), & a \leqq g(x) \leqq b, \\ a, & g(x) < a, \end{cases}$$

where $a \leqq 0$ and $b \geqq 0$. Then

$$\begin{aligned} s(x) &= \lim_{a\to-\infty}\lim_{b\to\infty}\lim_{N\to\infty} Q^N g_a^b(x) \\ &= \lim_{a\to-\infty}\lim_{N\to\infty}\lim_{b\to\infty} Q^N g_a^b(x) \\ &= \lim_{b\to\infty}\lim_{a\to-\infty}\lim_{N\to\infty} Q^N g_a^b(x). \end{aligned} \tag{2.53}$$

According to Remark 2 to Lemma 4,

$$\lim_{N\to\infty} G_a^N \varphi_a(x) = s_a(x) = \lim_{b\to\infty} \lim_{N\to\infty} Q^N g_a^b(x)$$
$$= \lim_{N\to\infty} \lim_{b\to\infty} Q^N g_a^b(x),$$

which, together with (2.52), proves the first two equalities in (2.53).

To prove the last inequality in (2.53) we let $s^\beta(x) = \sup_{\tau\in\mathfrak{R}} \mathbf{E}_x g^\beta(x_\tau)$, where

$$g^\beta(x) = \min\big(\beta, g(x)\big), \qquad \beta \geqq 0.$$

Then, by the second equality in (2.53),

$$s^\beta(x) = \lim_{a\to-\infty} \lim_{N\to\infty} Q^N g_a^\beta(x) \leqq s(x). \tag{2.54}$$

Since $s^\beta(x) \uparrow s(x)$ as $\beta \uparrow \infty$,

$$s^*(x) = \lim_{\beta\to\infty} s^\beta(x) = \lim_{\beta\to\infty} \lim_{a\to-\infty} \lim_{N\to\infty} Q^N g_a^\beta(x) \leqq s(x). \tag{2.55}$$

We establish the reverse inequality $s^*(x) \geqq s(x)$. Let $\tau \in \mathfrak{R}$. Then $\mathbf{E}_x g^-(x_\tau) < \infty$ for all $x \in E$, and since $g^b(x) \uparrow g(x)$ as $b \uparrow \infty$, by the Lebesgue monotone convergence theorem we get

$$\mathbf{E}_x g^b(x_\tau) \uparrow \mathbf{E}_x g(x_\tau) \qquad \text{as } b \uparrow \infty. \tag{2.56}$$

But $\mathbf{E}_x g^b(x_\tau) \leqq s^b(x) \leqq s^*(x)$, and consequently

$$s(x) = \sup_{\tau\in\mathfrak{R}} \mathbf{E}_x g(x_\tau) = \sup_{\tau\in\mathfrak{R}} \mathbf{E}_x g(x_\tau) \leqq s^*(x).$$

Thus

$$s(x) = s^*(x) = \lim_{b\to\infty} \lim_{a\to-\infty} \lim_{N\to\infty} Q^N g_a^b(x).$$

REMARK 3. We note that in the second equality in (2.53) the limits on a and N cannot, generally speaking, be interchanged, i.e.

$$s(x) = \lim_{a\to-\infty} \lim_{N\to\infty} \lim_{b\to\infty} Q^N g_a^b(x) \neq \lim_{N\to\infty} \lim_{a\to-\infty} \lim_{b\to\infty} Q^N g_a^b(x).$$

Indeed, in the example given at the beginning of this section,

$$Q^N g_a^b(x) = Q^N g_a(x), \quad \lim_{N\to\infty} Q^N g_a(x) = 0, \quad \lim_{a\to-\infty} Q^N g_a(x) = -x$$

and consequently

$$0 = \lim_{a\to-\infty} \lim_{N\to\infty} Q^N g_a(x) \neq \lim_{N\to\infty} \lim_{a\to-\infty} Q^N g_a(x) = -x$$

for all $x \neq 0$.

The theorem is proved.

REMARK 4. The function $s(x)$ satisfies the recursion equation

$$s(x) = \max\{g(x), Ts(x)\}.$$

This fact follows in an obvious fashion from the equality $s(x) = s_*(x)$ and equation (2.46).

3. PROOF OF THEOREM 3. Let

$$\begin{aligned} &g \in L, \quad b \geqq 0, \quad g^b(x) = \min(b, g(x)) \\ &s^b(x) = \sup_{\tau \in \mathfrak{R}} \mathbf{E}_x g^b(x_\tau) = \sup_{\tau \in \mathfrak{R}} \mathbf{E}_x g^b(x_\tau), \\ &\bar{s}^b(x) = \sup_{\tau \in \mathfrak{R}} \mathbf{E}_x g^b(x_\tau) = \sup_{\tau \in \bar{\mathfrak{R}}} \mathbf{E}_x g^b(x_\tau). \end{aligned}$$

Obviously $s^b(x) \leqq s(x)$, $\lim_{b\to\infty} s^b(x)$ exists and $s^*(x) = \lim_{b\to\infty} s^b(x) \leqq s(x)$. On the other hand, since $\tilde{\mathbf{E}}_x g^-(x_\tau) < \infty$ for $\tau \in \bar{\mathfrak{R}}$ and $g^b(x) \uparrow g(x)$ as $b \uparrow \infty$, it follows that $\tilde{\mathbf{E}}_x g^b(x_\tau) \uparrow \tilde{\mathbf{E}}_x g(x_\tau)$ as $b \uparrow \infty$. Hence

$$\tilde{\mathbf{E}}_x g^b(x_\tau) \leqq \bar{s}^b(x_\tau) = s^b(x) \leqq s^*(x), \quad \tilde{\mathbf{E}}_x g(x_\tau) \leqq s^*(x)$$

and

$$\bar{s}(x) = \sup_{\tau \in \bar{\mathfrak{R}}} \tilde{\mathbf{E}}_x g(x_\tau) \leqq s^*(x). \tag{2.57}$$

From a comparison of (2.57) with the established inequalities $s^*(x) \leqq s(x) \leqq \bar{s}(x)$ we obtain

$$s^*(x) = \lim_{b\to\infty} s^b(x) = \lim_{b\to\infty} \bar{s}^b(x) = \bar{s}(x) = s(x). \tag{2.58}$$

Since $\tilde{\mathbf{E}}_x \bar{s}^-(x_\tau) \leqq \tilde{\mathbf{E}}_x g^-(x_\tau) < \infty$ for $\tau \in \bar{\mathfrak{R}}$ and, by Lemma 10,

$$\bar{s}(x) \geqq \bar{s}^b(x) \geqq \tilde{\mathbf{E}}_x \bar{s}^b(x_\tau), \tag{2.59}$$

$\bar{s}^b(x) \uparrow \bar{s}(x)$ as $b \uparrow \infty$, it follows from (2.59) that $\bar{s}(x)$ is a regular excessive majorant of the function $g(x)$.

If $f(x)$ is a regular excessive majorant of $g(x)$, then $\tilde{\mathbf{E}}_x f(x_\tau)$ is defined for $\tau \in \bar{\mathfrak{R}}$ and

$$f(x) \geqq \tilde{\mathbf{E}}_x f(x_\tau) \geqq \tilde{\mathbf{E}}_x g(x_\tau),$$

whence $f(x) \geqq \bar{s}(x)$. Consequently $\bar{s}(x)$ is the smallest regular excessive majorant of the function $g(x)$.

Finally, since $s(x)$ is an excessive majorant of the function $g(x)$, we have $s(x) \geqq \max\{g(x), Ts(x)\}$. On the other hand, since $s^b(x) = \max\{g^b(x), Ts^b(x)\}$ and $s^b(x) \uparrow s(x)$ as $b \uparrow \infty$, we have $s(x) \leqq \max\{g(x), Ts(x)\}$. Thus

$$s(x) = \max\{g(x), Ts(x)\}.$$

REMARK 1. If $g \in L$, then

$$s(x) = \lim_{b\to\infty} \lim_{a\to-\infty} \lim_{N\to\infty} Q^N g_a^b(x), \tag{2.60}$$

which follows from (2.58) and Remark 2 to Lemma 10.

REMARK 2. Suppose the time $\tau^* \in \bar{\mathfrak{N}}$ ($\tau^* \in \mathfrak{N}$) is such that the associated reward $\tilde{f}(x) = \tilde{\mathbf{E}}_x g(x_{\tau^*})$ $\left(f(x) = \mathbf{E}_x g(x_{\tau^*})\right)$ is a regular excessive function and $\tilde{f}(x) \geqq g(x)$ $\left(f(x) \geqq g(x)\right)$. Then τ^* is a $(0,\bar{s})$-optimal time $\left((0,s)\text{-optimal time}\right)$.

REMARK 3. If $g \in L$, then Corollary 2 of Theorem 1 remains valid (with the obvious changes in notation).

REMARK 4. Theorem 3 and Theorem 4 (to be proved below) are also valid for terminating Markov processes. It is easy to make the necessary changes in notation and statement if we make use of the remarks made in § 2.5.

4. In this subsection we consider the problems of existence and structure of ε-optimal, $\varepsilon > 0$, Markov times. Theorem 4, proved below, is a direct generalization of the result contained in Theorem 2.

THEOREM 4. *Suppose the function $g \in L$ satisfies the condition*

$$A^+ \colon \ \mathbf{E}_x\Big[\sup_n g^+(x_n)\Big] < \infty$$

and $v(x)$ is its smallest $\bar{\mathfrak{N}}$-regular excessive majorant.

1) *For every $\varepsilon > 0$ the time*

$$\tau_\varepsilon = \inf\{n\colon v(x_n) \leqq g(x_n) + \varepsilon\}$$

is an ε-optimal stopping time (more precisely, $(0,s)$-optimal s.t.).

2) *The time*

$$\tau_0 = \inf\{n\colon v(x_n) = g(x_n)\}$$

is a $(0,\bar{s})$-optimal Markov time, i.e.

$$\tilde{\mathbf{E}}_x g(x_{\tau_0}) = \bar{s}(x) = v(x), \qquad x \in E.$$

3) *If the time τ_0 is a stopping time ($\tau_0 \in \mathfrak{M}$), then it is optimal (more precisely, $(0,s)$-optimal).*

PROOF. 1) Since $\mathbf{P}_x(\tau_\varepsilon < \infty) = 1$ for $x \in E$ under the hypothesis A^+, it follows that $\tau_\varepsilon \in \mathfrak{M}$. Then (ε,s)-optimality, $\varepsilon > 0$, of the time τ_ε was proved in Lemma 10 $\left(\text{inequalities (2.49) and (2.50)}\right)$.

2) To prove $(0,\bar{s})$-optimality of the time τ_0, we make use of Lemma 5. Due to the condition A^+, this lemma can be applied to $f(x) = v(x)$ and $\tilde{\mathbf{E}}_x v(x_{\tau_0})$ is defined. Thus for all $N \in \boldsymbol{N}$

$$\begin{aligned} v(x) &= \int_{(\tau_0<N)} v(x_{\tau_0})d\mathbf{P}_x + \int_{(\tau_0\geqq N)} v(x_N)d\mathbf{P}_x \\ &\leqq \int_{(\tau_0<N)} v(x_{\tau_0})d\mathbf{P}_x + \int_{(\tau_0\geqq N)} \mathbf{E}_{x_N}\left[\sup_n g(x_n)\right]d\mathbf{P}_x \\ &= \int_{(\tau_0<N)} v(x_{\tau_0})d\mathbf{P}_x + \int_{(\tau_0\geqq N)} \sup_{n\geqq N} g(x_n)d\mathbf{P}_x \\ &= \int_{(\tau_0<\infty)} \chi_{(\tau_0<N)}v(x_{\tau_0})d\mathbf{P}_x + \int_{\Omega} \chi_{(\tau_0\geqq N)} \sup_{n\geqq N} g(x_n)d\mathbf{P}_x. \end{aligned} \tag{2.61}$$

Hence, taking into account that $g\in L(A^+)$ and applying Fatou's Lemma, we obtain

$$\begin{aligned} v(x) &\leqq \limsup_N \int_{(\tau_0<\infty)} \chi_{(\tau_0<N)}v(x_{\tau_0})d\mathbf{P}_x + \limsup_N \int_{\Omega} \chi_{(\tau_0\geqq N)} \sup_{n\geqq N} g(x_n)d\mathbf{P}_x \\ &\leqq \int_{(\tau_0<\infty)} v(x_{\tau_0})d\mathbf{P}_x + \int_{(\tau_0=\infty)} \limsup_N g(x_N)d\mathbf{P}_x \\ &= \int_{(\tau_0<\infty)} g(x_{\tau_0})d\mathbf{P}_x + \int_{(\tau_0=\infty)} \limsup_N \ g(x_N)d\mathbf{P}_x = \tilde{\mathbf{E}}_x g(x_{\tau_0}). \end{aligned} \tag{2.62}$$

According to the condition A^+ we have $\tau_0 \in \bar{\mathfrak{N}}$, and therefore it follows from (2.62) that $v(x) = \tilde{s}(x) = \tilde{\mathbf{E}}_x g(x_{\tau_0})$.

3) From (2.61) and Lemma 10 we have

$$v(x) \leqq \int_{(\tau_0<N)} v(x_{\tau_0})d\mathbf{P}_x + \int_{(\tau_0\geqq N)} \sup_n g^+(x_n)d\mathbf{P}_x,$$

whence

$$v(x) \leqq \int_{(\tau_0<\infty)} v(x_{\tau_0})d\mathbf{P}_x = \int_{\tau_0<\infty} g(x_{\tau_0})d\mathbf{P}_x = \mathbf{E}_x g(x_{\tau_0}),$$

which proves the $(0,s)$-optimality of the stopping time τ_0.

COROLLARY. *If the condition A^+ is satisfied and* $\lim_n g(x_n) = -\infty$ ($\mathbf{P}_x$-*a.s.*, $x\in E$), *then the time τ_0 is $(0,s)$-optimal.*

Indeed, if $\mathbf{P}_{x_0}(\tau_0 = \infty) > 0$ for some x_0, then $\tilde{s}(x_0) = -\infty$, which contradicts the inequalities $\tilde{s}(x_0) \geqq g(x_0) > -\infty$. Therefore the time $\tau_0 \in \mathfrak{M}$, which fact, together with the condition A^+, proves the $(0,s)$-optimality of this time.

§ 4. Optimal stopping rules in the class $\mathfrak{M}_N$

1. Suppose $\mathfrak{M}_N \subseteq \mathfrak{M}$ is the class of stopping times τ such that $\mathbf{P}_x(\tau \leqq N) = 1$, $N < \infty$, for all $x \in E$. We let L_N denote the set of $\mathscr{B}$-measurable functions $-\infty < g(x) \leqq \infty$ satisfying the condition

$$\mathbf{E}_x g^-(x_n) < \infty, \qquad n = 0, 1,\ldots, N, \tag{2.63}$$

and set

$$s_N(x) = \sup_{\tau\in\mathfrak{M}_N} \mathbf{E}_x g(x_\tau). \tag{2.64}$$

In the preceding sections we studied the properties of the values $s(x)$ and $\bar{s}(x)$ under the assumption that the given Markov times belong to the classes $\mathfrak{M}$ and $\overline{\mathfrak{M}}$. It is now natural to raise the question of the structure of the values $s_N(x)$, optimal stopping rules in the class $\mathfrak{M}_N$ (they exist in the case $N < \infty$, as we shall see later), and convergence $s_N(x) \to s(x)$ as $N \to \infty$. Clearly, problems of optimal stopping in which the duration of observation is bounded by N lead to consideration of the values $s_N(x)$.

The basic results for the case $N < \infty$ are stated in the following theorem, in which we use the notation

$$s_n(x) = \sup_{\tau \in \mathfrak{M}_n} \mathbf{E}_x g(x_\tau), \qquad n \leqq N.$$

THEOREM 5. *Suppose the function $g \in L_N$. Then for all $1 \leqq n \leqq N$*

$$s_n(x) = Q^n g(x) \tag{2.65}$$

$$s_n(x) = \max\{g(x), Ts_{n-1}(x)\}, \quad s_0(x) = g(x), \tag{2.66}$$

and the time

$$\tau_N = \min\{n \leqq N: s_{N-n}(x_n) = g(x_n)\} \tag{2.67}$$

is optimal, i.e. for all $x \in E$

$$\mathbf{E}_x g(x_{\tau_N}) = \sup_{\tau \in \mathfrak{M}_N} \mathbf{E}_x g(x_\tau).$$

PROOF. Let $g_n(x) = Q^n g(x)$ and $g_0(x) = g(x)$. By Remark 3 to Lemma 4 (§ 1),

$$g_n(x) = \max\{g(x), Tg_{n-1}(x)\}, \qquad 1 \leqq n \leqq N. \tag{2.68}$$

For each $n \leqq N$ we let

$$\xi_m^{(n)}(\omega) = g_{n-m}(x_m(\omega)), \qquad m \leqq n. \tag{2.69}$$

Then it is easy to see that the system

$$\{\xi_m^{(n)}(\omega), \mathscr{F}_m, \mathbf{P}_x\}, \quad m \leqq n, \quad \mathscr{F}_m = \sigma\{\omega: x_0, \ldots, x_m\}$$

forms a (generalized) super-martingale for each $x \in E$:

$$\mathbf{E}_x(\xi_m^{(n)})^- < \infty, \quad \mathbf{E}_x[\xi_{m+1}^{(n)} \mid \mathscr{F}_m] \leqq \xi_m^{(n)} \qquad (\mathbf{P}_x\text{-a. s., } x \in E).$$

To prove the assertions (2.65) and (2.66), it is sufficient to show that $s_n(x) = g_n(x)$, $1 \leqq n \leqq N$.

To this end, for each $1 \leqq n \leqq N$ we define the Markov time

$$\sigma_n = \min\{i: g_{n-i}(x_i) = g(x_i)\}, \qquad 0 \leqq i \leqq n, \tag{2.70}$$

and show that for each $\tau \in \mathfrak{M}_n$

$$\mathbf{E}_x g(x_{\sigma_n}) \geqq \mathbf{E}_x g(x_\tau), \qquad x \in E, \tag{2.71}$$

and

$$\mathbf{E}_x g(x_{\sigma_n}) = g_n(x), \qquad x \in E. \tag{2.72}$$

We now obtain the desired equality $s_n(x) = g_n(x)$, $1 \leqq n \leqq N$, in an obvious fashion from (2.71) and (2.72). In order to prove (2.71) we shall establish the validity of the following chain of inequalities:

$$\mathbf{E}_x g(x_{\sigma_n}) = \mathbf{E}_x g_{n-\sigma_n}(x_{\sigma_n}) \geqq \mathbf{E}_x g_{n-\tau}(x_\tau) \geqq \mathbf{E}_x g(x_\tau). \tag{2.73}$$

The last inequality is obvious, since $g_k(x) \geqq g(x)$, $k \geqq 0$. Further, by (2.70),

$$\begin{aligned} \mathbf{E}_x g(x_{\sigma_n}) = \mathbf{E}_x g_{n-\sigma_n}(x_{\sigma_n}) &= \int_{(\sigma_n \geqq \tau)} g_{n-\sigma_n}(x_{\sigma_n}) d\mathbf{P}_x + \int_{(\sigma_n < \tau)} g_{n-\sigma_n}(x_{\sigma_n}) d\mathbf{P}_x \\ &= \sum_{i=0}^{n} \int_{(\sigma_n \geqq i,\ \tau = i)} g_{n-\sigma_n}(x_{\sigma_n}) d\mathbf{P}_x + \int_{(\sigma_n < \tau)} g_{n-\sigma_n}(x_{\sigma_n}) d\mathbf{P}_x. \end{aligned} \tag{2.74}$$

From Theorem 5 of Chapter I, for any two Markov times τ_1 and τ_2, $\tau_2 \geqq \tau_1$ ($\mathbf{P}_x$-a.s., $x \in E$), belonging to the class $\mathfrak{M}_n$ we have

$$\mathbf{E}_x[\xi_{\tau_2}^{(n)}(\omega) \mid \mathscr{F}_{\tau_1}] \leqq \xi_{\tau_1}^{(n)}(\omega) \qquad (\mathbf{P}_x\text{-a. s., } x \in E). \tag{2.75}$$

Since the event $\{\sigma_n < \tau\} \in \mathscr{F}_{\sigma_n}$ (see § 2 of Chapter I), by (2.75) the last integral in (2.74) is equal to

$$\int_{(\sigma_n < \tau)} g_{n-\sigma_n}(x_{\sigma_n}) d\mathbf{P}_x = \int_{(\sigma_n < \tau)} \xi_{\sigma_n}^{(n)} d\mathbf{P}_x \geqq \int_{(\sigma_n < \tau)} \xi_{\tau}^{(n)} d\mathbf{P}_x = \int_{(\sigma_n < \tau)} g_{n-\tau}(x_\tau) d\mathbf{P}_x. \tag{2.76}$$

Now we shall show that for all $0 \leqq i \leqq n$ (see (2.74))

$$\int_{(\sigma_n \geqq i,\ \tau = i)} g_{n-\sigma_n}(x_{\sigma_n}) d\mathbf{P}_x = \int_{(\sigma_n \geqq i,\ \tau = i)} g_{n-i}(x_i) d\mathbf{P}_x. \tag{2.77}$$

Indeed, on the set $\{\omega : \sigma_n > i\}$

$$g_n(x_0(\omega)) > g(x_0(\omega)), \ldots, g_{n-i}(x_i(\omega)) > g(x_i(\omega))$$

and consequently, according to (2.68),

$$g_n(x_0(\omega)) = Tg_{n-1}(x_0(\omega)), \ldots, g_{n-i}(x_i(\omega)) = Tg_{(n-(i+1))}(x_i(\omega)).$$

Therefore

$$\begin{aligned} \int_{(\sigma_n \geqq i, \tau = i)} g_{n-i}(x_i) d\mathbf{P}_x &= \int_{(\sigma_n = i, \tau = i)} g_{n-i}(x_i) d\mathbf{P}_x + \int_{(\sigma_n > i, \tau = i)} g_{n-i}(x_i) d\mathbf{P}_x \\ &= \int_{(\sigma_n = i, \tau = i)} g_{n-i}(x_i) d\mathbf{P}_x + \int_{(\sigma_n > i, \tau = i)} Tg_{n-(i+1)}(x_i) d\mathbf{P}_x \\ &= \int_{(\sigma_n = i, \tau = i)} g_{n-\sigma_n}(x_{\sigma_n}) d\mathbf{P}_x + \int_{(\sigma_n > i, \tau = i)} g_{n-(i+1)}(x_{i+1}) d\mathbf{P}_x \\ &= \int_{(i \leqq \sigma_n \leqq i+1, \tau = i)} g_{n-\sigma_n}(x_{\sigma_n}) d\mathbf{P}_x = \ldots \\ &= \int_{(i \leqq \sigma_n \leqq n, \tau = i)} g_{n-\sigma_n}(x_{\sigma_n}) d\mathbf{P} + \int_{(\sigma_n > i+1, \tau = i)} g_{n-(i+1)}(x_{i+1}) d\mathbf{P}_x, \end{aligned}$$

which proves (2.77).

From (2.74), (2.76) and (2.77) we obtain

$$\begin{aligned}\mathbf{E}_x g(x_{\sigma_n}) &= \sum_{i=0}^{n} \int_{(\sigma_n \geqq i, \tau = i)} g_{n-\sigma_n}(x_{\sigma_n}) d\mathbf{P}_x + \int_{(\sigma_n < \tau)} g_{n-\sigma_n}(x_{\sigma_n}) d\mathbf{P}_x \\ &\geqq \sum_{i=0}^{n} \int_{(\sigma_n \geqq i, \tau = i)} g_{n-i}(x_i) d\mathbf{P}_x + \int_{(\sigma_n < \tau)} g_{n-\tau}(x_\tau) d\mathbf{P}_x \\ &= \mathbf{E}_x g_{n-\tau}(x_\tau) \geqq \mathbf{E}_x g(x_\tau).\end{aligned}$$

Thus we have proved the chain of inequalities (2.73), from which, in particular, it follows that the time σ_n is optimal in the class $\mathfrak{M}_n$:

$$s_n(x) = \mathbf{E}_x g(x_{\sigma_n}). \tag{2.78}$$

We now establish the equality (2.72). From (2.73) for $\tau \equiv 0$ we obtain

$$s_n(x) \geqq \mathbf{E}_x g_n(x_0(\omega)) = g_n(x).$$

On the other hand, in view of (2.70) and (2.75),

$$\begin{aligned}s_n(x) &= \mathbf{E}_x g(x_{\sigma_n}) = \mathbf{E}_x g_{n-\sigma_n}(x_{\sigma_n}) \\ &= \mathbf{E}_x \xi^{(n)}_{\sigma_n}(\omega) \leqq \mathbf{E}_x \xi^{(n)}_0(\omega) = \mathbf{E}_x g_n(x_0(\omega)) = g_n(x),\end{aligned}$$

which proves the equalities (2.72) and (2.66). Obviously $\sigma_N = \tau_N$. Consequently the time τ_N, defined in (2.67), is optimal.

2. As follows from the theorem just proved, in the case $N < \infty$ the values $s_1(x), \cdots, s_N(x)$ can be found successively by the formula $s_n(x) = Q^n g(x)$. Knowing these values also makes it possible (at least in principle) to find the optimal time

$$\tau_N = \min\{n : g(x_n) = s_{N-n}(x_n)\}.$$

Now suppose that $N \to \infty$. Since $s_{N+1}(x) \geqq s_N(x)$ and $\tau_{N+1}(\omega) \geqq \tau_N(\omega)$ ($\mathbf{P}_x$-a. s., $x \in E$), the following limits exist:

$$s^*(x) = \lim_{N\to\infty} s_N(x) \quad \text{and} \quad \tau^*(\omega) = \lim_{N\to\infty} \tau_N(\omega).$$

It is natural to ask the question of when does $s^*(x)$ coincide with the value $s(x) = \sup_{\tau \in \mathfrak{M}} \mathbf{E}_x g(x_\tau)$ and whether the time τ^* is $(0,s)$-optimal or $(0,\bar{s})$-optimal.

THEOREM 6. 1) *If the function $g \in L(A^-)$, then*

$$s^*(x) = \lim_{N\to\infty} s_N(x) = s(x) = \bar{s}(x). \tag{2.79}$$

2) *If the function $g \in L(A^-, A^+)$, then the time $\tau^* = \lim_{N\to\infty} \tau_N$ is $(0,\bar{s})$-optimal.*

3) *If the function $g \in L(A^-, A^+)$ and the time $\tau^* \in \mathfrak{M}$, then it is $(0,s)$-optimal.*

4) *The time $\tau^* = \lim_{N\to\infty} \tau_N = \inf\{n : g(x_n) = s(x_n)\}$.*

PROOF. 1) The assertion (2.79) follows from Lemma 4 and Theorem 1, since $s_N(x) = Q^N(x)$.

2) On the set $\{\tau^* < \infty\}$

$$\lim_{N\to\infty} g(x_{\tau_N}(\omega)) = g(x_{\tau^*}(\omega)).$$

Therefore, by Fatou's Lemma and (2.79),

$$\begin{aligned}\tilde{s}(x) &= \lim_{N\to\infty} s_N(x) \\ &= \limsup_N \mathbf{E}_x g(x_{\tau_N}) \leqq \mathbf{E}_x\left[\limsup_N g(x_{\tau_N})\right] \\ &= \int_{(\tau^*<\infty)} \limsup_N g(x_{\tau_N}) d\mathbf{P}_x + \int_{(\tau^*=\infty)} \limsup_N g(x_{\tau_N}) d\mathbf{P}_x \\ &= \int_{(\tau^*<\infty)} g(x_{\tau^*}(\omega)) d\mathbf{P}_x + \int_{(\tau^*=\infty)} \limsup_n g(x_n) d\mathbf{P}_x = \tilde{\mathbf{E}}_x g(x_{\tau^*}),\end{aligned} \tag{2.80}$$

whence follows the $(0,\tilde{s})$-optimality of the time τ^*.

3) If $\mathbf{P}_x(\tau^* = \infty) = 0$, $x \in E$, then from (2.80) we obtain that the time τ^* is $(0,s)$-optimal.

4) To prove the last assertion of the theorem, we set

$$\tilde{\tau} = \inf\{n : g(x_n) = s(x_n)\}$$

and show that $\tilde{\tau} = \tau^*$ ($\mathbf{P}_x$-a. s., $x \in E$).

Suppose $\omega_0 \in \{\tilde{\tau} = n\}$, $n < \infty$. Then $g(x_i) < s_{N-i}(x_i)$, $i = 0,\cdots,n-1$, where $x_i = x_i(\omega_0)$, and consequently $g(x_i) < s(x_i)$ for sufficiently large N, i.e. $\tau_N(\omega_0) \geqq n$. Therefore $\tau^*(\omega_0) \geqq \tau_N(\omega_0) \geqq \tilde{\tau}(\omega_0)$. If $\tilde{\tau}(\omega_0) = \infty$, then $g(x_i) < s(x_i)$ for $x_i = x_i(\omega_0)$ for all $i \geqq 0$, and consequently $\tau_N(\omega_0) > n$ for sufficiently large N and $n < N$, whence $\tau^*(\omega_0) = \lim_{N\to\infty} \tau_N(\omega_0) > n$, i.e. $\tau^*(\omega_0) = \infty$.

Conversely, suppose $\omega_0 \in \{\tau^* = n\}$, $n < \infty$. Then there is a sufficiently large N such that $\tau_N(\omega_0) = n$ and thus $g(x_i) < s_N(x_i)$, $i = 0, 1,\cdots, n-1$, whence $g(x_i) < s(x_i)$, i.e., $\tilde{\tau}(\omega_0) \geqq n$.

If $\tau^*(\omega_0) = \infty$ and $\tilde{\tau}(\omega_0) = k$, then $g(x_i) < s(x_i)$, $i = 0,1,\cdots,k-1$, $g(x_k) = s(x_k)$ and $x_i = x_i(\omega_0)$, and consequently, for all sufficiently large N, $g(x_i) < s_N(x_i)$, $i = 0,1,\cdots,k-1$, and $g(x_k) = s_N(x_k)$, i.e. $\tau_N(\omega_0) = k$, which contradicts the assumption that $\tau^*(\omega_0) = \lim_{N\to\infty} \tau_N(\omega_0) = \infty$.

REMARK. From the proof of Theorem 6 it is evident that assertions 2 and 3 of the theorem can also be stated in the following way.

Suppose $g \in L(A^-)$ and $\mathbf{E}_{x^0}[\sup_n g^+(x_n)] < \infty$ for given $x_0 \in E$. Then the time $\tau^* = \lim_{N\to\infty} \tau_N$ is $(0,\tilde{s})$-optimal at the point x_0, i.e. $\tilde{s}(x_0) = \tilde{\mathbf{E}}_{x_0} g(x_{\tau^*})$. If also $\mathbf{P}_{x_0}(\tau^* < \infty)$, then the time τ^* is $(0,s)$-optimal at the point x_0.

3. Application of part 3 of Theorem 6 is difficult because it is not easy to verify that the Markov time τ^* belongs to the class $\mathfrak{M}$. However, sometimes general considerations make it possible to establish that an optimal time σ^* exists in the class $\mathfrak{M}$. It turns out that under this assumption we have $(0,s)$-optimality of the time $\tau^* = \lim_{N\to\infty} \tau_N$, and the time τ^* is the smallest among all the $(0,s)$-optimal times.

THEOREM 7. *Suppose the function $g \in L(A^-, A^+)$, and an optimal stopping time σ^* exists. Then the time $\tau^* \leqq \sigma^*$ ($\mathbf{P}_x$-a.s., $x \in E$) is also optimal in the class $\mathfrak{M}$.*

PROOF. It suffices to establish that $\tau^* \leqq \sigma^*$, since then the (0,s)-optimality of the time τ^* follows from the assumption $\sigma^* \in \mathfrak{M}$ and Theorem 6.

We let $A_n = \{\omega : \sigma^*(\omega) = n\}$, $n \in \boldsymbol{N}$, and show that on this set $s(x_n) = g(x_n)$ ($\mathbf{P}_x$-a.s., $x \in E$). The desired inequality $\tau^*(\omega) \leqq \sigma^*(\omega)$ will then follow.

Suppose $\omega \in A_n$ and $\mathfrak{M}^{(n)} \subseteq \mathfrak{M}$ is the class of stopping times such that $\mathbf{P}_x(\tau \geqq n) = 1$ for $x \in E$. Then for each $\tau \in \mathfrak{M}^{(n)}$ on the set A_n

$$g(x_n) \geqq \mathbf{E}_x[g(x_\tau) \mid \mathscr{F}_n] \qquad (\mathbf{P}_x\text{-a.s., } x \in E). \tag{2.81}$$

Indeed, suppose (2.81) is not satisfied for some $x \in E$. Then the set

$$B = A_n \cap \{\omega : g(x_n) < \mathbf{E}_x\,[g(x_\tau) \mid \mathscr{F}_n]\}$$

has positive $\mathbf{P}_x$-measure.

We form the time

$$\bar{\tau}(\omega) = \chi_B(\omega)\,\tau(\omega) + \chi_{\bar{B}}(\omega)\,\sigma^*(\omega). \tag{2.82}$$

It is obvious that $\bar{\tau}(\omega)$ takes on its values in the set $\boldsymbol{N}$. We shall show that $\bar{\tau}(\omega)$ is a Markov time.

If $k \geqq n$, then[12]

$$\begin{aligned}\{\omega : \bar{\tau}(\omega) = k\} &= \{\omega : \bar{\tau}(\omega) = k\} \cap B + \{\omega : \bar{\tau}(\omega) = k\} \cap B \\ &= \{\omega : \tau(\omega) = k\} \cap B + \{\omega : \sigma^*(\omega) = k\} \cap B \in \mathscr{F}_k.\end{aligned}$$

Suppose $k < n$. In the case $k = 0$

$$\begin{aligned}\{\chi_B(\omega)\,\tau(\omega) + \chi_{\bar{B}}(\omega)\,\sigma^*(\omega) = 0\} &= \{\chi_B\tau = 0\} \cap \{\chi_B\sigma^* = 0\} \\ &= \{\chi_B\tau = 0\} \cap \{\sigma^* = 0\} \cap \{\chi_{\bar{B}} = 1\} \\ &\quad + \{\chi_B\tau = 0\} \cap \{\sigma^* = 0\} \cap \{\chi_{\bar{B}} = 0\} \\ &\quad + \{\chi_B\tau = 0\} \cap \{\sigma^* \neq 0\} \cap \{\chi_{\bar{B}} = 0\} \\ &= \{\sigma^* = 0\} + \phi + \{\tau = 0\} \cap \{\sigma^* = n\}.\end{aligned}$$

The set $\{\tau = 0\} \in \mathscr{F}_0$, and $\mathbf{P}_x\{\tau = 0\} = 0$ for all $x \in E$. Since we are assuming[13] that $\mathscr{F}_0 = \bar{\mathscr{F}}_0$, by the definition of $\bar{\mathscr{F}}_0$ the set $\{\tau = 0\} \cap \{\sigma^* = n\}$ also belongs to $\bar{\mathscr{F}}_0$, since $\phi \subseteq \{\tau = 0\} \cap \{\sigma^* = n\} \subseteq \{\tau = 0\}$ and $\mathbf{P}_x\{\tau = 0\} = 0$, $x \in E$. Consequently $\{\omega : \bar{\tau}(\omega) = 0\} \in \bar{\mathscr{F}}_0 = \mathscr{F}_0$.

Now suppose $0 < k < n$. Then

---

[12] $C_1 + C_2$ denotes the set-theoretic sum of the sets C_1 and C_2 if $C_1 \cap C_2 = \phi$.
[13] See footnote 1 on p. 27.

$$\begin{aligned}
&\{\chi_B\tau + \chi_{\bar{B}}\sigma^* = k\} \\
&\quad = \{\chi_B\tau = k\} \cap \{\chi_{\bar{B}}\sigma^* = 0\} + \{\chi_B\tau = 0\} \cap \{\chi_{\bar{B}}\sigma^* = k\} \\
&\quad = \{\chi_B = 1\} \cap \{\tau = k\} + \{\chi_{\bar{B}} = 1\} \cap \{\sigma^* = k\} \\
&\quad = \{\chi_B = 1\} \cap \{\tau = k\} + \{\sigma^* = k\}.
\end{aligned}$$

The set $\{\tau = k\} \in \mathscr{F}_k$, and $\mathbf{P}_x\{\tau = k\} = 0$, $x \in E$, $k < n$. Thereofre $\{\chi_B = 1\} \cap \{\tau = k\} \in \bar{\mathscr{F}}_k = \mathscr{F}_k$, and, since $\{\sigma^* = k\} \in \mathscr{F}_k$, it follows that $\{\bar{\tau} = k\} \in \mathscr{F}_k$ for $k < n$.

Thus $\bar{\tau} = \bar{\tau}(\omega)$ is a Markov time, and $\bar{\tau} \in \mathfrak{M}$ For this time, according to (2.82), we have

$$\begin{aligned}
\mathbf{E}_x g(x_{\bar{\tau}}) &= \mathbf{E}_x\{\chi_B g(x_\tau) + \chi_{\bar{B}} g(x_\sigma)\} \\
&= \mathbf{E}_x\{\chi_B \mathbf{E}_x [g(x_\tau) \mid \mathscr{F}_n] + \chi_{\bar{B}} g(x_{\sigma^*})\} \\
&> \mathbf{E}_x\{\chi_B g(x_n) + \chi_{\bar{B}} g(x_{\sigma^*})\} \\
&= \mathbf{E}_x\{\chi_B g(x_{\sigma^*}) + \chi_{\bar{B}} g(x_{\sigma^*})\} = \mathbf{E}_x g(x_{\sigma^*}),
\end{aligned}$$

which contradicts the optimality of the time σ^*.

Thus (2.81) is satisfied on the set A_n for all $\tau \in \mathfrak{M}^{(n)}$. We let $\hat{\mathfrak{M}}^{(n)} \subseteq \mathfrak{M}^{(n)}$ denote the class of Markov times $\hat{\tau}^n$ which are the first entry times (after the time n) of the set $A \in \mathscr{B}$, i.e.

$$\hat{\tau}^n = \inf\{t \geqq n : x_t \in A\}.$$

Then, in view of the Markov property of the given process X and the fact that $\theta_n x_{\hat{\tau}_0} = x_{\hat{\tau}^n}$ ([25], Chapter 4, property 4.1. E), we obtain

$$\begin{aligned}
\mathbf{E}_x[g(x_{\tau^n}) \mid \mathscr{F}_n] &= \mathbf{E}_x[g(\theta_n x_{\hat{\tau}^0}) \mid \mathscr{F}_n] \\
&= \mathbf{E}_x[\theta_n g(x_{\hat{\tau}^0}) \mid \mathscr{F}_n] = \mathbf{E}_{x_n}[g(x_{\hat{\tau}^0})].
\end{aligned}$$

Therefore it follows from (2.81) that for all $\hat{\tau}^0 \in \hat{\mathfrak{M}}^{(0)}$

$$g(x_n) \geqq \mathbf{E}_{x_n}[g(x_{\hat{\tau}^0})] \qquad (\mathbf{P}_x\text{-a.s., } x \in E). \tag{2.83}$$

But according to Corollary 2 of Theorem 1 ($\hat{\mathfrak{M}}^{(0)} = \hat{\mathfrak{M}}$)we have

$$s(x) = \sup_{\hat{\tau} \in \hat{\mathfrak{M}}} \mathbf{E}_x g(x_{\hat{\tau}}) \tag{2.84}$$

Comparing (2.83) and (2.84), we obtain that on A_n

$$g(x_n) \geqq s(x_n) \qquad (\mathbf{P}_x\text{-a.s., } x \in E).$$

Since we always have $s(x_n) \geqq g(x_n)$, consequently $s(x_n(\omega)) = g(x_n(\omega))$ on the set $A_n = \{\omega : \sigma^*(\omega) = n\}$, whence $\tau^*(\omega) \leqq n$, which proves the inequality $\tau^*(\omega) \leqq \sigma^*(\omega)$.

§ 5. On uniqueness of the solution of the recursion equations

1. From Theorem 3 it follows that the value $s(x) = \sup_{\tau \in \mathfrak{M}} \mathbf{E}_x g(x_\tau)$, $g \in L$, satisfies the recursion equation

$$f(x) = \max\{g(x),\ Tf(x)\}. \tag{2.85}$$

Therefore if this equation has a unique solution (in the class L), then it coincides with the value. Unfortunately, even if equation (2.85) does have a solution, as a rule this solution is not unique. For example, if $0 \leqq g(x) \leqq C < \infty$ and $P(x,E) = 1$, $x \in E$, then any constant $K \geqq C$ is a solution of the given equation.

With regard to this, it is important to study the problem of uniqueness of solutions of equation (2.85) under various assumptions about the classes of "admissible" functions $f = f(x)$.

2. Suppose

$$P(n,x,\Gamma) = \mathbf{P}_x\{x_n \in \Gamma\}, \quad \mu(n,x,\Gamma) = \frac{1}{n}\sum_{i=1}^{n} P(i,x,\Gamma), \quad \Gamma \in \mathscr{B}, \quad n \in \boldsymbol{N}.$$

Assume that there exists a nonnegative measure $\mu(\cdot)$ on $(E,\mathscr{B})$ such that for every $\mathscr{B}$-measurable bounded function $f = f(x)$, $x \in E$, we have

$$\int_E f(y)\,\mu(n,x,dy) \to \int_E f(y)\,\mu(dy), \qquad n \to \infty,$$

for all $x \in E$.

THEOREM 8. *Suppose $f_1(x)$ and $f_2(x)$ are two solutions of equation* (2.85) *belonging to the class L and coinciding on some measurable set $\Lambda \subseteq E$, and such that*

$$\sup_{x\in E} |f_1(x) - f_2(x)| < \infty.$$

If $\mu(E \setminus \Lambda) < 1$, then $f_1(x) \equiv f_2(x)$.

PROOF. Let $r(x) = |f_1(x) - f_2(x)|$. Then from (2.85) it can easily be established that

$$\begin{aligned} |f_1(x) - f_2(x)| &= |\max\{g(x), Tf_1(x)\} - \max\{g(x), Tf_2(x)\}| \\ &\leqq |Tf_1(x) - Tf_2(x)| \leqq T|f_1 - f_2|(x), \end{aligned} \tag{2.86}$$

i.e. $r(x) \leqq Tr(x)$, whence $r(x) \leqq T^n r(x)$, and thus

$$r(x) \leqq \int_E r(y)\,\mu(n,x,dy). \tag{2.87}$$

Passing to the limit in (2.87) as $n \to \infty$, we find

$$r(x) \leqq \int_E r(y)\,\mu(dy) \leqq \sup_{y\in E} r(y)\,\mu(E \setminus \Lambda)$$

and

$$\sup_{x\in E} r(x) \leqq \sup_{x\in E} r(x) \cdot \mu(E \setminus \Lambda).$$

Since by hypothesis $\mu(E \setminus \Lambda) < 1$, it follows that $r(x) \equiv 0$, i.e $f_1(x) \equiv f_2(x)$.

COROLLARY 1. *If $P(x,E) = p < 1$ for all $x \in E$, then the solution of equation* (2.85) *is unique in the class of measurable bounded functions.*

COROLLARY 2. *If the function $g(x)$ is bounded in modulus and $f(x)$ is a bounded solution of equation* (2.85), *and coincides with $g(x)$ on the set Λ, where $\xi(E \setminus \Lambda) <$ 1, then $f(x)$ is the smallest excessive majorant of the function $g(x)$, and consequently $f(x) = s(x)$.*

For the proof it is sufficient to note that the value $s(x)$ also satisfies equation (2.85) and $f(x)$ coincides with $s(x)$ on Λ.

3. Another criterion for the coincidence of two solutions of equation (2.85) is given by

THEOREM 9. *Suppose $f_1(x)$ and $f_2(x)$ are two measurable solutions of equation* (2.85) *such that*

$$\mathbf{E}_x\left\{\sup_n |f_1(x_n) - f_2(x_n)|\right\} < \infty, \qquad x \in E.$$

Suppose that for each $\varepsilon > 0$ there exists a set $\Lambda_\varepsilon \in \mathscr{B}$ such that

1) $|f_1(x) - f_2(x)| < \varepsilon$, $x \in \Lambda_\varepsilon$;

2) $\mathbf{P}_x\{x_n \in \Lambda_\varepsilon$ *for infinitely many* $n \in \boldsymbol{N}\} = 1$, $x \in E$.

Then $f_1(x) \equiv f_2(x)$.

PROOF. We form the process

$$R = \big(r(x_n), \mathscr{F}_n, \mathbf{P}_x\big), \qquad n \in \boldsymbol{N}, \qquad x \in E.$$

Since $0 \leqq r(x) \leqq T^n r(x)$ according to (2.86), it follows that the process R is a non-negative sub-martingale. From the condition $\sup_{n \in N} \mathbf{E}_x r(x_n) < \infty$, $x \in E$, and Theorem 3 of Chapter I, it follows that the limit $\lim_{n\to\infty} r(x_n)$ exists with $\mathbf{P}_x$-probability one. By condition 2), $x_n \in \Lambda_\varepsilon$ for infinitely many $n \in \boldsymbol{N}$, and therefore, since $\varepsilon >$ 0 is arbitrary, $\lim_{n\to\infty} r(x_n) = 0$ ($\mathbf{P}_x$-a.s., $x \in E$). Then from the inequality

$$0 \leqq r(x) \leqq \mathbf{E}_x r(x_n)$$

and Fatou's Lemma we obtain

$$0 \leqq r(x) \leqq \limsup_{n\to\infty} \mathbf{E}_x r(x_n) \leqq \mathbf{E}_x \limsup_{n\to\infty} r(x_n) = 0,$$

which proves the theorem.

COROLLARY 1. *Suppose $\Lambda = \bigcap_{\varepsilon>0} \Lambda_\varepsilon$, i.e.*

$$\Lambda = \{x: |f_1(x) - f_2(x)| = 0\},$$

and $\mathbf{P}_x\{x_n \in \Lambda$ *for infinitely many* $n \in \boldsymbol{N}\} = 1$. *Then, under the hypotheses of Theorem* 9, $f_1(x) \equiv f_2(x)$.

COROLLARY 2. *If the solution* $f(x)$ *of equation* (2.85) *coincides with the function* $g(x)$ *on a set* Λ *such that* $\mathbf{P}_x\{x_n \in \Lambda$ *for infinitely many* $n \in \boldsymbol{N}\} = 1$, *where* $\mathbf{E}_x\{\sup_n |f(x_n)|\} < \infty$ *and* $\mathbf{E}_x\{\sup_n |g(x_n)|\} < \infty$, *then* $f(x)$ *coincides with the value* $s(x)$.

4. In concluding, we give one more useful theorem which makes it possible for us to judge whether a given solution of equation (2.85) is a value.

THEOREM 10. *Suppose* $g \in L(A^-,A^+)$, *and the function* $f \in L(A^+)$ *is a solution of equation* (2.85). *Now if*

$$\limsup_n g(x_n) = \limsup_n f(x_n), \tag{2.88}$$

then $f(x) \equiv s(x)$.

PROOF. For $\varepsilon > 0$ let

$$\tau_\varepsilon = \inf\{n: g(x_n) \geqq f(x_n) - \varepsilon\}.$$

Then, in view of (2.88) and the condition $g \in L(A^+)$ (see the proof of Lemma 9), we have $\mathbf{P}_x(\tau_\varepsilon < \infty) = 1$, $x \in E$, whence by Lemma 5

$$s(x) \geqq \mathbf{E}_x g(x_{\tau_\varepsilon}) \geqq \mathbf{E}_x f(x_{\tau_\varepsilon}) - \varepsilon = f(x) - \varepsilon.$$

Since $\varepsilon > 0$ is arbitrary, $s(x) \geqq f(x)$. But if $g \in L(A^-)$, then by Theorem 1 the value $s(x)$ is the s.e.m. of the function $g(x)$. Consequently $s(x) \equiv f(x)$.

§ 6. Criteria for "boundedness" of optimal stopping rules

1. Suppose $\mathfrak{M} = \{\tau\}$ is the class of Markov stopping times, and $\mathbf{P}_x(\tau < \infty) = 1$, for all $x \in E$. Assume there exists an optimal stopping rule $\tau^* \in \mathfrak{M}$, $\mathbf{E}_x g(x_{\tau^*}) = s(x)$, $x \in E$. Here it may be the case that for some state $x \in E$ there is a finite $N(x)$ such that $\mathbf{P}_x(\tau^* \leqq N(x)) = 1$. In this case we say that the optimal stopping rule τ^* is "bounded" at the point x. If there is a finite N such that $\mathbf{P}_x(\tau^* \leqq N) = 1$ for all $x \in E$, in other words if $\tau^* \in \mathfrak{M}_N$, then the stopping rule τ^* is said to be "bounded."

In this section we investigate a number of criteria which make it possible to decide for which initial states $x \in E$ optimal stopping rules are "bounded."

We also show criteria which allow us to determine if a given stopping time bound $N(x)$ is exact [i.e. $\mathbf{P}_x(\tau^* \leqq N(x)) = 1$ and $\mathbf{P}_x(\tau^* = N(x)) > 0$].

2. Everywhere below we assume that the function $g \in L(A^-)$. According to Theorem 1 and Lemma 4,

$$s(x) = \lim_{N\to\infty} Q^N g(x). \tag{2.89}$$

Let

$$s_k(x) = \sup_{\tau \in \mathfrak{M}_k} \mathbf{E}_x g(x_\tau) \qquad (= Q^k g(x)), \tag{2.90}$$

$$\alpha_k(x) = Q^k g(x) - TQ^k g(x), \tag{2.91}$$

$$\beta_k(x) = s_{k+1}(x) - s_k(x). \tag{2.92}$$

According to (2.92) and (2.66), for all $n \geqq k$

$$\beta_k(x_{n-k}) = \max\{g(x_{n-k}), TQ^k g(x_{n-k})\} - Q^k g(x_{n-k}) \qquad (\mathbf{P}_x\text{-a.s., } x \in E). \tag{2.93}$$

Here it is clear from (2.91) and (2.93) that the condition $\beta_k(x_{n-k}) = 0$ is equivalent to the condition $\alpha_k(x_{n-k}) \geqq 0$.

THEOREM 11. *If for a given state $x \in E$ for some $k \geqq 0$ there exists a finite $n_k = n_k(x)$ such that with $\mathbf{P}_x$-probability one*

$$\beta_k(x_{n-k}) = 0, \qquad n \geqq n_k, \tag{2.94}$$

then $s_n(x) = s(x)$ for all $n \geqq n_k$.

If (2.94) *is satisfied for k and l, $l < k$, and $N_k = N_k(x)$ and $N_l = N_l(x)$ are the smallest of the numbers $n_k = n_k(x)$ and $n_l = n_l(x)$ satisfying* (2.94), *then $N_k \leqq N_l$.*

For the proof of the theorem we need

LEMMA 11. *For all $n \geqq 0$ and $x \in E$*

$$s_{n+2}(x) - s_{n+1}(x) \leqq T(s_{n+1} - s_n)(x). \tag{2.95}$$

The proof is based on an analysis of the recursion relations

$$s_{k+1}(x) = \max\{g(x), Ts_k(x)\}, \qquad k \geqq 0, \tag{2.96}$$

and reduces to an investigation of the following three cases.

a) If $g(x) \geqq Ts_{n+1}(x)$, then $g(x) \geqq Ts_{n+1}(x) \geqq Ts_n(x)$, and from (2.96) for $k = n + 1$ and $k = n$ we obtain $s_{n+2}(x) = g(x)$, $s_{n+1}(x) = g(x)$. Consequently (2.95) is satisfied.

b) If $g(x) \leqq Ts_n(x)$, then $g(x) \leqq Ts_n(x) \leqq Ts_{n+1}(x)$, and from (2.96) we get $s_{n+1}(x) = Ts_n(x)$ and $s_{n+2}(x) = Ts_{n+1}(x)$. Consequently $s_{n+2}(x) - s_{n+1}(x) = T(s_{n+1} - s_n)(x)$.

c) If $Ts_n(x) \leqq g(x) \leqq Ts_{n+1}(x)$, then

$$s_{n+2}(x) - s_{n+1}(x) = Ts_{n+1}(x) - g(x) \leqq Ts_{n+1}(x) - Ts_n(x) = T(s_{n+1} - s_n)(x),$$

which proves (2.95).

PROOF OF THEOREM 11. By (2.95),

$$\begin{aligned} 0 &\leqq s_{n_k+1}(x) - s_{n_k}(x) \\ &\leqq T(s_{n_k} - s_{n_k-1})(x) \leqq \dots \leqq T^{n_k-k}(s_{k+1} - s_k)(x) \\ &= \mathbf{E}_x[s_{k+1}(x_{n_k-k}) - s_k(x_{n_k-k})]. \end{aligned}$$

But $0 \leqq s_{k+1}(x_{n_k-k}) - s_k(x_{n_k-k}) = \beta_k(x_{n_k-k})$, and therefore if $\beta_k(x_{n_k-k}) = 0$ with $\mathbf{P}_x$-probability one (or, equivalently, $\alpha_k(x_{n_k-k}) \geqq 0$), then $s_{n_k+1}(x) = s_{n_k}(x)$. Analogously $s_n(x) = s_{n_k}(x)$ for all $n \geqq n_k$, whence $s_n(x) = s(x)$ for $n \geqq n_k$.

The second assertion of the theorem follows from the inequality $N_k \leqq N_{k-1}$, which can easily be derived from (2.95).

COROLLARY 1. *If for given $x \in E$ and some $k \geqq 0$ there exists a finite $n_k = n_k(x)$ such that $\beta_k(x_{n-k}) = 0$ with $\mathbf{P}_x$-probability one for $n \geqq n_k$, then the stopping time*

$$\tau_{n_k} = \min\{m: s_{n_k-m}(x_m) = g(x_m)\}$$

is optimal (at the given point x):

$$s(x) = \mathbf{E}_x g(x_{\tau_{n_k}}).$$

COROLLARY 2. *If $M_k = \sup_x n_k(x) < \infty$ for some $k \geqq 0$, then the stopping time*

$$\tau_{M_k} = \min\{m: S_{M_k-m}(x_m) = g(x_m)\}$$

is optimal:

$$\mathbf{E}_x g(x_{\tau_{M_k}}) = s(x) \qquad \textit{for all} \quad x \in E.$$

REMARK. Of course it is simplest to construct criteria for "boundedness" for small k. Thus for $k = 0$

$$\alpha_0(x_n) = g(x_n) - Tg(x_n);$$

for $k = 1$

$$\alpha_1(x_{n-1}) = Qg(x_{n-1}) - TQg(x_{n-1}).$$

Hence for $k = 0$, if there is an $n_0 < \infty$ such that for all $x \in E$ we have $g(x_n) \geqq Tg(x_n)$, $n \geqq n_0$, with $\mathbf{P}_x$-probability one, then an optimal stopping rule τ^* automatically exists and $\mathbf{P}_x(\tau^* \leqq n_0) = 1$ for all $x \in E$.

According to the second part of Theorem 11, $N_0(x) \geqq N_1(x)$. Therefore the criterion based on analyzing the variable $\alpha_1(x_{n-1})$ yields a more precise estimate from above for the stopping time bound: $N(x) \leqq N_1(x) \leqq N_0(x)$.

Since $N(x) \leqq N_k(x)$, it is of interest to clarify when $N(x) = N_k(x)$ for some k and all or some $x \in E$.

THEOREM 12. *If for given $x \in E$ and some $k \geqq 0$ there is a finite $N_k = N_k(x)$ such that with $\mathbf{P}_x$-probability one*

$$\beta_k(x_{n-k}) = 0, \qquad n \geqq N_k, \tag{2.97}$$

and the following inequalities are satisfied with positive $\mathbf{P}_x$-probability:

$$Tg(x_i) \geqq g(x_i), \quad i = 0, 1, \ldots, N_k - k - 2, \qquad \beta_k(x_{N_k-k-1}) > 0, \tag{2.98}$$

then

$$s_{N_k-1}(x) < s_{N_k}(x) = s_{N_k+1} = \dots = s(x) \tag{2.99}$$

and $N(x) = N_k(x)$.

We prove one lemma as a preliminary.

LEMMA 12. *If for given* $x \in E$ *in the space* E^{n-m}, $n - m \geqq 1$, *there exists a set* $A_1 \times \dots \times A_{n-m}$ *of positive* $\mathbf{P}_x$*-probability such that*

$$Tg(x_i) \geqq g(x_i), \quad i = 0, 1, \dots, n - m - 1, \quad x_i \in A_i,$$
$$\beta_m(x_{n-m}) > 0, \quad x_{n-m} \in A_{n-m},$$

then

$$\beta_{n-i}(x_i) > 0, \quad i = 0, 1, \dots, n - m - 1, \quad x_i \in A_i. \tag{2.100}$$

The proof is carried out by induction. Suppose the inequality (2.100) is satisfied for $i = j + 1, \dots, n - m - 1$. We establish its validity for $i = j \geqq 0$.

If $x_k \in A_k$, $0 \leqq k \leqq n - m - 1$, then

$$g(x_k) \leqq Tg(x_k) \leqq Ts_{n-k-1}(x_k) \leqq Ts_{n-k}(x_k).$$

Hence for the given $x \in E$ we have

$$\begin{aligned}\beta_{n-j}(x_j) &= s_{n-j+1}(x_j) - s_{n-j}(x_j) = T(s_{n-j} - s_{n-j-1})(x_j) \\ &= \mathbf{E}_x[s_{n-j}(x_{j+1}) - s_{n-j-1}(x_{j+1}) \mid x_j] \\ &\geqq \mathbf{E}_x\{\chi_{A_{j+1}}(x_{j+1})[s_{n-1}(x_{j+1}) - s_{n-j-1}(x_{j+1})] \mid x_j\} \\ &= \mathbf{E}_x\{\chi_{A_{j+1}}(x_{j+1})\beta_{n-j-1}(x_{j+1}) \mid x_j\} > 0 \qquad (\mathbf{P}_x\text{-a. s.}).\end{aligned}$$

The lemma is proved.

PROOF OF THEOREM 12. We note that $\beta_{N_k-1}(x_0) = s_{N_k}(x_0) - s_{N_k-1}(x_0)$ for $x_0 = x$. Therefore, if we take $i = 0$ and $n = N_k$ in (2.100), then we obtain $\beta_{N_k-1}(x) > 0$, i.e. $s_{N_k-1}(x) < s_{N_k}(x)$.

By the preceding theorem $s_{N_k}(x) = s_{N_k+1}(x) = \dots = s(x)$, and thus (2.99) is proved. From (2.99) and Corollary 1 of Theorem 11 it also follows that $N_k(x) = N(x)$.

§ 7. Sufficient and randomized classes of stopping times[14)]

1. The solution of every concrete problem in mathematical statistics is usually started by seeking "sufficient statistics," i.e. data from past observations which contain all the "essential information" needed for determining the "best solution."

Also well known is the role played by "randomized decision rules" in mathematical statistics.

---

14) We restrict consideration of these concepts to stopping times only. They can be derived and investigated in a completely analogous manner for Markov times.

In this section we consider the concepts of "sufficient" and "randomized" classes of Markov times in connection with the problem of optimal stopping of Markov processes with discrete time.

2. Let $(\Omega,\mathscr{F})$ be a measurable space, and $X = (x_n,\mathscr{F}_n,\mathbf{P}_x)$ a Markov chain, $n \in \boldsymbol{N}$, with values in the phase space $(E,\mathscr{B})$, where $\mathscr{F}_n \subseteq \mathscr{F}_{n+1}, \mathscr{F}_n \subseteq \mathscr{F}$ and the measures $\mathbf{P}_x$ are given on the smallest σ-algebra containing $\mathscr{F}_n$ for all $n \in \boldsymbol{N}$.

Suppose that in $\mathscr{F}$ we distinguish the system $F^* = \{\mathscr{F}_n^*\}$, $n \in \boldsymbol{N}$, of σ-algebras $\mathscr{F}_n^*$ having the properties $\mathscr{F}_n^* \subseteq \mathscr{F}_{n+1}^*$ and $\mathscr{F}_n \subseteq \mathscr{F}_n^*$.

We also assume that on the smallest σ-algebra containing $\mathscr{F}_n^*$ for all $n \in \boldsymbol{N}$ we are given the probability measures $\mathbf{P}_x^*$ which are extensions of the measures $\mathbf{P}_x$ (i.e. $\mathbf{P}_x^*(A) = \mathbf{P}_x(A)$ if $A \in \sigma(\bigcup_{n\in N} \mathscr{F}_n)$), and the process $X^* = (x_n, \mathscr{F}_n^*,\mathbf{P}_x^*)$ is Markov.

We let $\mathfrak{M}(F)$ and $\mathfrak{M}(F^*)$ denote the classes of stopping times such that $\{\tau = n\} \in \mathscr{F}_n$ and $\{\tau = n\} \in \mathscr{F}_n^*$, $n \in \boldsymbol{N}$, respectively. Obviously $\mathfrak{M}(F^*) \supseteq \mathfrak{M}(F)$.

DEFINITION 1. A stopping time $\tau \in \mathfrak{M}(F^*)$ is said to be *randomized* with respect to the system of σ-algebras $F = \{\mathscr{F}_n\}$, $n \in \boldsymbol{N}$.

Let

$$s^*(x) = \sup_{\tau\in\mathfrak{M}(F)} \mathbf{E}_x^* g(x_\tau) \quad\text{and}\quad s(x) = \sup_{\tau\in\mathfrak{M}(F)} \mathbf{E}_x g(x_\tau),$$

where $\mathbf{E}_x^*$ denotes averaging with respect to the measure $\mathbf{P}_x^*$ and $g \in L(A^-)$. Clearly $s^*(x) \geqq s(x)$. However, by the same token $s^*(x) = s(x)$. Indeed, according to Corollary 2 of Theorem 1, the values $s^*(x)$ and $s(x)$ coincide with $\hat{s}(x) = \sup_{\tau\in\hat{\mathfrak{M}}} \mathbf{E}_x g(x_\tau)$, where $\hat{\mathfrak{M}}$ is the class of first entry times of the sets $A \in \mathscr{B}$, $\hat{\mathfrak{M}} \subseteq \mathfrak{M}(F) \subseteq \mathfrak{M}(F^*)$. To prove the equality $s^*(x) = s(x)$ we can also make use of Lemma 4 and Theorem 1.

Indeed, if

$$g \in L(A^-), \qquad \left(\mathbf{E}_x^*\left[\sup_n g^-(x_n)\right] = \mathbf{E}_x\left[\sup_n g^-(x_n)\right] < \infty, \quad x \in E\right),$$

then

$$s^*(x) = \lim_{N\to\infty} Q^{*N} g(x), \qquad s(x) = \lim_{N\to\infty} Q^N g(x),$$

where $Q^*g(x) = \max\{g(x), \mathbf{E}_x^*g(x_1)\}$. But $\mathbf{E}_x^*g(x_1) = \mathbf{E}_x g(x_1)$, since $\mathbf{P}_x^*(A) = \mathbf{P}_x(A)$, $A \in \sigma(\bigcup_n \mathscr{F}_n)$. Therefore $Q^*g(x) = Qg(x)$ and, analogously, $Q^{*N}g(x) = Q^N g(x)$ for all $N \in \boldsymbol{N}$.

Thus if $g \in L(A^-)$, then $s^*(x) = s(x)$. In the general case we use Remark 1 of Theorem 3 for the proof ($g \in L$).

Thus we have proved

THEOREM 13. *Suppose the function $g \in L$. Then $s^*(x) = s(x)$, i.e. the additional introduction of randomized stopping times does not increase the value.*

Although it follows from this theorem that randomization does not increase the value, we can nevertheless demonstrate a number of valuable applications of randomized stopping times.

For example, if the value $s(x) = \infty$ for some $x \in E$, then there may not be an optimal time in the classes $\mathfrak{M}$ and $\mathfrak{M}(F)$ (and also $\overline{\mathfrak{M}}(F) = \overline{\mathfrak{M}}$) while at the same time there is such a time in the class $\mathfrak{M}(F^*)$.

Indeed, suppose $s(x) = \infty$ for some x. Then there is a sequence of stopping times $\{\tau_i\}$, $i \in \boldsymbol{N}$, $\tau_i \in \mathfrak{M}(F)$, such that $s(x) = \sup_i \mathbf{E}_x g(x_{\tau_i})$. Without loss of generality we can assume that $\mathbf{E}_x g(x_{\tau_i}) \geqq 2^i$.

Let $\xi(\omega)$ be an $\mathscr{F}$-measurable random variable taking on the values $i = 1,2,\cdots$ with probabilities 2^{-i}, where[15)]

$$\mathbf{P}_x^*\{[\xi(\omega) = i] \cap A\} = \mathbf{P}_x(A)\cdot 2^{-i}$$

for all $x \in E$ and $A \in \sigma(\bigcup_{n \in \boldsymbol{N}} \mathscr{F}_n)$. We define the randomized time $\tau^* = \tau^*(\omega)$ such that $\tau^*(\omega) = \tau_i(\omega)$ if $\xi(\omega) = i$. Then, obviously,

$$\mathbf{E}_x^* g(x_{\tau^*}) = \sum_{i=1}^{\infty} \mathbf{E}_x g(x_{\tau_i})\cdot 2^{-i} = \infty.$$

An investigation of randomized stopping times is especially valuable for solving variational problems on optimal stopping. For example, suppose we want to find $\sup \mathbf{E}_{x_0} g(x_\tau)$, $x_0 \in E$, under the assumption that we are considering only those stopping times τ for which $\mathbf{E}_{x_0} f(x_\tau) = c$, where c is a constant and $f, g \in L(A^-)$. Even in those cases where there exist τ_1 and τ_2 belonging to $\mathfrak{M}(F)$ and such that $\mathbf{E}_{x_0} f(x_{\tau_1}) = a < c$ and $\mathbf{E}_{x_0} f(x_{\tau_2}) = b > c$, we can generally not find a time τ in the class $\mathfrak{M}(F)$ for which $\mathbf{E}_{x_0} f(x_\tau) = c$. However, in the class $\mathfrak{M}(F^*)$ the time

$$\tau^*(\omega) = \tau_i(\omega) \quad \text{if} \quad \xi(\omega) = i,$$

where $i = 1,2$ and $\xi(\omega)$ is an $\mathscr{F}$-measurable random variable such that

$$\mathbf{P}_x^*(\xi(\omega) = 1) = \frac{b-c}{b-a}, \quad \mathbf{P}_x^*(\xi(\omega) = 2) = \frac{c-a}{b-a},$$
$$\mathbf{P}_x^*\{[\xi(\omega) = i] \cap A\} = \mathbf{P}_x(A) \cdot \mathbf{P}_x^*(\xi(\omega) = i),$$

yields $\mathbf{E}_{x_0} f(x_{\tau^*}) = c$.

3. We now turn to the concept of sufficient classes of stopping times.

Suppose $X = (x_n, \mathscr{F}_n, \mathbf{P}_x)$ is a Markov process, $n \in \boldsymbol{N}$, $x \in E$ and $s(x) = \sup_{\tau \in \mathfrak{M}(F)} \mathbf{E}_x g(x_\tau)$, where $F = \{\mathscr{F}_n\}$, $n \in \boldsymbol{N}$.

DEFINTION 2. The class of stopping times $\mathfrak{M}(G)$, where $G = \{\mathscr{G}_n\}$, $n \in \boldsymbol{N}$, and

---

15) We are assuming that the original space $(\Omega, \mathscr{F})$ is sufficiently "rich." Otherwise, instead of $(\Omega, \mathscr{F})$, we would have to consider the new space $(\tilde{\Omega}, \tilde{\mathscr{F}})$, where $\tilde{\Omega} = \Omega \times \Omega^*$, $\tilde{\mathscr{F}} = \mathscr{F} \times \mathscr{F}^*$ and $(\Omega^*, \mathscr{F}^*)$ is some measurable space of "randomized" outcomes $\omega^* \in \Omega^*$.

the σ-algebras $\mathscr{G}_n$ are such that $\mathscr{G}_n \subseteq \mathscr{G}_{n+1}$ and $\mathscr{G}_n \subseteq \mathscr{F}_n$, is said to be *sufficient* if for all $x \in E$

$$\sup_{\tau \in \mathfrak{M}(F)} \mathbf{E}_x g(x_\tau) = \sup_{\tau \in \mathfrak{M}(G)} \mathbf{E}_x g(x_\tau). \tag{2.101}$$

Thus, if randomization has led to a broadening of the class $\mathfrak{M}(F)$, then, conversely, sufficiency is introduced with the aim of narrowing the class of stopping times without decreasing the value.

It follows from Theorem 1 that the class of stopping times $\mathfrak{M}(G)$, where $G = \{\mathscr{G}_n\}$ and $\mathscr{G}_n = \sigma\{\omega: x_0, x_1, \ldots, x_n\}$, $n \in \boldsymbol{N}$, is sufficient since $\mathfrak{M}(G) \supseteq \hat{\mathfrak{M}}$ and

$$\sup_{\tau \in \mathfrak{M}(F)} \mathbf{E}_x g(x_\tau) = \sup_{\tau \in \hat{\mathfrak{M}}} \mathbf{E}_x g(x_\tau), \qquad g \in L(A^-).$$

In this regard it is valuable to note that if the process $X = (x_n, \mathscr{F}_n, \mathbf{P}_x)$ is Markov, then the process $\tilde{X} = (x_n, \mathscr{F}_n, \mathbf{P}_x)$, $n \in \boldsymbol{N}$, is also Markov. Hence in view of the sufficiency of the class $\mathfrak{M}(G)$ it follows that we can consider the process $\tilde{X}$ directly in solving optimal stopping problems.

Is it impossible to carry out a further narrowing of the class $\mathfrak{M}(G)$ ut whtio decreasing the value? From this point of view the simplest class is of course the class of stopping times τ identically equal to some moment of time n, $\tau(\omega) \equiv n$, $n \in \boldsymbol{N}$. It is obvious that this class coincides with the class $\mathfrak{M}(G^0)$, where $G^0 = \{\mathscr{G}_n^0\}$, $n \in \boldsymbol{N}$, and every σ-algebra $\mathscr{G}_n^0$ is trivial, i.e. $\mathscr{G}_n^0 = \{\phi, \Omega\}$, where ϕ is the empty set.

However, even though nontrivial examples exist (see the example given below) in which the class $\mathfrak{M}(G^0)$ is sufficient, nevertheless these cases are exceptions rather than the rule.

EXAMPLE. Suppose $\xi, \xi_1, \xi_2, \cdots$ is a sequence of independent identically distributed random variables such that $m = \mathbf{E}e^{\lambda\xi} < \infty$, where λ is a constant. Let $x_n = x + \xi_1 + \cdots + \xi_n$, $x \in R$. Then the sequence (n, x_n), $x_0 = x$, forms a homogeneous Markov chain. Let $g(n,x) = c(n)e^{\lambda x}$, $c(n) \geqq 0$ and $s(n,x) = \sup \mathbf{E}_x g(\tau, x_\tau)$, where the upper bound is taken over all stopping times greater than or equal to n.

It is easy to compute that

$$Q^N g(n,x) = e^{\lambda x} \max_{0 \leqq k \leqq N} [m^k c(n+k)].$$

Therefore

$$s(n,x) = e^{\lambda x} \max_{k \geqq 0} [m^k c(n+k)]. \tag{2.102}$$

If $\mathbf{E}_x[\sup_{n \geqq 0} c(n) e^{\lambda x_n}] < \infty$, then according to Theorem 2 the time

$$\tau_\varepsilon = \inf\{n: s(n,x_n) \leqq g(n, x_n) + \varepsilon\}, \qquad \varepsilon > 0,$$

is an ε-optimal stopping time. By (2.102)

$$\tau_\varepsilon = \inf\left\{n: \max_{k \geq 0} [m^k c(n+k)] \leq c(n) + \varepsilon e^{-\lambda x_n}\right\}.$$

Therefore if there exists a finite N such that

$$\max_{k \geq 0} [m^k c(N+k)] = c(N),$$

then τ_0 is a nonrandom optimal stopping time equal to the first value of n for which

$$c(n) = \max_{k \geq 0} [m^k c(n+k)].$$

4. The theorem given below concerning a class of sufficient stopping times will be used many times later for solving a number of concrete problems.

Suppose $X = (x_n, \mathscr{F}_n, \mathbf{P}_x)$ and $\varphi = \varphi(x)$ is a measurable mapping of $(E, \mathscr{B})$ into some measurable space $(E_\varphi, \mathscr{B}_\varphi)$.

THEOREM 14. *Let* $g = g(\varphi(x)) \in L$. *If for every* $A \in \mathscr{B}_\varphi$

$$\mathbf{P}_x\{\varphi(x_1) \in A\} = f_A(\varphi(x)), \tag{2.103}$$

where $f_A(\varphi)$ *is a* $\mathscr{B}_\varphi$*-measurable function, then the class* $\mathfrak{N}(\Phi)$, *where*

$$\Phi = \{\Phi_n\},\ n \in \boldsymbol{N} \quad \textit{and} \quad \Phi_n = \sigma\{\omega: \varphi(x_0), \dots, \varphi(x_n)\},$$

is sufficient:

$$s(x) = \sup_{x \in \mathfrak{N}(F)} \mathbf{E}_x g(\varphi(x_\tau)) = \sup_{\tau \in \mathfrak{N}(\Phi)} \mathbf{E}_x(\varphi g(x_\tau)). \tag{2.104}$$

PROOF. If $g \in L(A^-, A^+)$, then $s(x) = \lim_{N \to \infty} Q^N g(\varphi(x))$, and the time

$$\tau_\varepsilon = \inf\{n: s(x_n) \leq g(\varphi(x_n)) + \varepsilon\}, \quad \varepsilon > 0,$$

is ε-optimal. But, by (2.103), for all N

$$Q^N g(\varphi(x)) = F_N(\varphi(x)),$$

where $F_N = F_N(\varphi)$ is a $\mathscr{B}_\varphi$-measurable function. Therefore $s(x) = F(\varphi(x))$, where the function $F(\varphi)$ is also $\mathscr{B}_\varphi$-measurable, and consequently $\tau_\varepsilon \in \mathfrak{N}(\Phi)$, which proves the theorem in the case $g \in L(A^-, A^+)$. To prove the general case we have to make use of Remark 1 to Theorem 3.

COROLLARY. *Suppose* $X = (Y, Z) = ((y_n, z_n), \mathscr{F}_n, \mathbf{P}_{y,z})$, $n \in \boldsymbol{N}$, *is a Markov process in the phase space* $(E_Y \times E_Z, \mathscr{B}_Y \times \mathscr{B}_Z)$, *and let* $\varphi(x) = z$ *and* $\mathbf{P}_{y,z}(z_1 \in A) = f_A(z)$. *Then the value*

$$s(y, z) = \sup_{\tau \in \mathfrak{N}(F)} \mathbf{E}_{y,z} g(z_\tau) = \sup_{\tau \in \mathfrak{N}(\Phi)} \mathbf{E}_{y,z} g(z_\tau),$$

where $\Phi = \{\Phi_n\}, \Phi_n = \sigma\{\omega\colon z_0, \ldots, z_n\}, n \in \boldsymbol{N}$. *The function* $s(y,z)$ *does not depend on* y (*more precisely, it is* $(\phi, E_Y) \times \mathscr{B}_z$*-measurable*).

§ 8. Optimal stopping of Markov sequences for functions $g(n,x)$ and in the presence of a fee

1. Let $X = (x_n, \mathscr{F}_n, \mathbf{P}_x)$, $n \in \boldsymbol{N}$, be a Markov chain in the phase space $(E, \mathscr{B})$. Up to now we have assumed that when we stopped the observation process in state x_n, we obtained a reward equal to $g(x_n)$.

We consider some generalizations of the given formulation of the optimal stopping problem. First we assume that the reward obtained in state x_n is given by the function $g(n, x_n)$ depending explicitly on n.

Suppose the function $g(n,x)$ is such that $\mathbf{E}_x g^-(n,x_n) < \infty$ for all n. Let $\mathfrak{N} \subseteq \mathfrak{M}$ denote the class of stopping times τ for which $\mathbf{E}_x g^-(\tau,x_\tau) < \infty$, and let

$$s(0, x) = \sup_{x \in \mathfrak{N}} \mathbf{E}_x g(\tau, x_\tau). \tag{2.105}$$

In order to reduce the problem of finding the value $s(0,x)$ and of optimal stopping times to the case already examined in which g did not depend on n, we form the homogeneous Markov chain[16] $X' = (x'_n, \mathscr{F}'_n, \mathbf{P}'_x)$, $n \in \boldsymbol{N}$, where $x'_n = (n, x_n)$ and $\mathscr{F}'_n = \mathscr{F}_n$, and if $x' = (n,x)$ then $\mathbf{P}'_x(A) = \mathbf{P}'_x(A)$, $A \in \sigma(\bigcup_{n \in \mathfrak{N}} \mathscr{F}_n)$.

We use $\mathfrak{N}^{(n)} \subseteq \mathfrak{N}$ to denote the class of stopping times $\tau \in \mathfrak{N}$ for which $\mathbf{P}_x(\tau \geqq n) = 1$, $x \in E$, and let the expression $|x'| = n$ denote that $x' = (n, x)$, where $x \in E$.

Fro all $n \in \boldsymbol{N}$ and $x \in E$ we set

$$s(n, x) = \sup_{\tau \in \mathfrak{N}^n} \mathbf{E}_x g(\tau, x_\tau) \tag{2.106}$$

or, equivalently,

$$s(x') = \sup_{\tau \in \mathfrak{N}^{(|x'|)}} \mathbf{E}'_{x'} g(x'_\tau), \tag{2.107}$$

where $\mathbf{E}'_{x'}$ denotes the mean with respect to the measure $\mathbf{P}'_x$ and $\mathfrak{N}^{(0)} = \mathfrak{N}$.

We call $f(x')$ a regular excessive majorant of the function $g(x')$ if

$$f(x') \geqq g(x'), \qquad x' \in \boldsymbol{N} \times E,$$
$$f(x') \geqq \mathbf{E}'_{x'} f(x'_\tau), \quad x' \in \boldsymbol{N} \times E,$$

for each $\tau \in \mathfrak{N}$.

The proofs of Theorems 1—4, with almost no change, allow us to establish the following result.

THEOREM 15. *The value* $s(x')$ *is the smallest regular*[17] *excessive majorant of the function* $g(x')$.

---

[16] This approach was used before in considering the example in § 7.
[17] We cannot talk about "regularity" in the case $\mathbf{E}_x[\sup_n g^-(n,x_n)] < \infty$, $x \in E$.

If $\mathbf{E}_x[\sup_n g^+(n,x_n)] < \infty$, $x \in E$, *then for all* $x' = (n, x)$ *the time*

$$\tau_\varepsilon^{(n)} = \inf \{m \geqq n: s(m, x_m) \leqq g(m, x_m) + \varepsilon\}, \qquad \varepsilon > 0,$$

is ε*-optimal in the class* $\mathfrak{N}^{(n)}$. *If also* $\mathbf{P}_x(\tau_0^{(n)} < \infty) = 1$, $x \in E$, *then the time* $\tau_0^{(n)}$ *is an optimal stopping time.*

If $\hat{\mathfrak{R}}^{(n)} \subseteq \mathfrak{N}^{(n)}$ *is the class of stopping times* $\hat{\tau}^{(n)}$ *such that* $\hat{\tau}^{(n)} = \min \{m \geqq n: x_m \in A_m\}$, *where* $A_m \in \mathscr{B}$, *then*

$$\sup_{\tau \in \mathfrak{N}^{(|x'|)}} \mathbf{E}_{x'} g(x'_\tau) = \sup_{\tau \in \mathfrak{R}^{(|x'|)}} \mathbf{E}_{x'} g(x'_\tau) = \sup_{\tau \in \hat{\mathfrak{R}}^{(|x'|)}} \mathbf{E}_{x'} g(x'_\tau). \tag{2.108}$$

The functions $s(n,x)$ *satisfy the equations*

$$s(n, x) = \max \{g(n, x), \mathbf{E}_x s(n + 1, x_1)\} \tag{2.109}$$

and, under the condition $\mathbf{E}_x[\sup_n g^-(n, x_n)] < \infty$, $x \in E$,

$$s(n, x) = \lim_{N\to\infty} Q^N g(n, x), \tag{2.110}$$

where

$$Qg(n, x) = \max \{g(n, x), Tg(n, x)\},$$
$$Tf(n, x) = \mathbf{E}_x f(n + 1, x_1).$$

REMARK. Above we assumed that the Markov process $X=(x_n, \mathscr{F}_n, \mathbf{P}_x)$, $n \in \mathbf{N}$, is homogeneous. In the case of an inhomogeneous process X (see § 3 of Chapter I), optimal stopping problems can be reduced to those already considered if we go over to the homogeneous process X', $x'_n = (n, x_n)$.

2. Many statistical problems, such as problems of sequential testing of hypotheses, are such that the possibility of making the next observation implies a possible decrease in the total reward.

More precisely, we assume that having stopped observing at the moment of time n, we obtain a reward (which may indeed turn out to be negative) equal to

$$G(n, x_0 \ldots, x_n) = \alpha^n g(x_n) - \sum_{s=0}^{n-1} \alpha^s c(x_s), \qquad n \geqq 1, \tag{2.111}$$

and $G(0, x_0) = g(x_0)$ for $n = 0$. In (2.111), α is a constant, $0 < \alpha \leqq 1$, and the functions $g(x)$ and $c(x)$ from the class L are assumed to satisfy the condition

$$\mathbf{E}_x G^-(n, x_0, \ldots, x_n) < \infty, \quad n \geqq 0. \tag{2.112}$$

Thus $c(x_s)$ can be treated as the fee for the possibility of making the next observation, in x_s, state and α as the parameter accounting for the change in "worth" over time.

Value is the quantity

$$s(x) = \sup \mathbf{E}_x\left\{\alpha^\tau g(x_\tau) - \sum_{s=0}^{\tau-1} \alpha^s c(x_s)\right\}, \tag{2.113}$$

where the upper bound is taken over the class $\mathfrak{N} = \mathfrak{N}_{g,c}$ of all stopping times τ for which the mathematical expectations on the right side of (2.113) are defined (for all $x \in E$).

In order to give an "excessive" characterization of value and demonstrate a method for finding ε-optimal stopping times, we give a

DEFINITION. The function $f \in L$ is said to be (α, c) *-excessive* if

$$f(x) \geqq \alpha Tf(x) - c(x), \qquad x \in E, \tag{2.114}$$

and to be an (α, c) *-excessive majorant of the function* $g(x)$ if also $f(x) \geqq g(x)$.

The (α, c)-excessive function $f(x)$ is said to be *regular* if

$$f(x) \geqq \mathbf{E}_x\left[\alpha^\tau f(x_\tau) - \sum_{s=0}^{\tau-1} \alpha^s c(x_s)\right], \qquad x \in E,$$

for all

$$\tau \in \mathfrak{N}_{f,c} = \left\{\tau : \mathbf{E}_x\left[\alpha^\tau f(x_\tau) - \sum_{s=0}^{\tau-1} \alpha^s c(x_s)\right]^- < \infty, x \in E\right\}.$$

THEOREM 16. *The value $s(x)$ defined in* (2.113) *is the smallest regular*[18] (α, c)*-excessive majorant of the function $g(x)$ and satisfies the condition*

$$s(x) = \max\{g(x), \alpha Ts(x) - c(x)\}. \tag{2.115}$$

If $\mathbf{E}_x[\sup_n G^+(n, x_0, \ldots, x_n)] < \infty$, $x \in E$, *then the time*

$$\tau_\varepsilon = \inf\{n : \alpha^n s(x_n) \leqq \alpha^n g(x_n) + \varepsilon\}, \quad \varepsilon > 0,$$

is an ε-optimal stopping time.

If, moreover, τ_0 is a stopping time, then it is optimal.

If $\mathbf{E}_x[\sup_n G^-(n, x_0, \ldots, x_n)] < \infty$, $x \in E$, *then*

$$s(x) = \lim_{N\to\infty} Q^N_{(\alpha,c)} g(x), \tag{2.116}$$

where $Q_{(\alpha,c)} g(x) = \max\{g(x), \alpha Tg(x) - c(x)\}$.

PROOF. Let $(E_n^*, \mathscr{B}_n^*)$ denote the phase space of the sequences

$$x_n^* = (n, x_0, \ldots, x_n), \quad x_i \in E,$$

$$E^* = \bigcup_{n=0}^{\infty} E_n^*, \qquad \mathscr{B}^* = \sigma\left(\bigcup_{n=0}^{\infty} \mathscr{B}_n^*\right).$$

[18] If $\mathbf{E}_x[\sup_n G^-(n, x_0, \ldots, x_n)] < \infty$, $x \in E$, then the function $s(x)$ is automatically regular.

If $x^* \in E^*$, then the expression $|x^*| = n$ also denotes that $x^* \in E^*_n$. Suppose $E^*_{m,x,}$, where $|x^*| = n \leqq m$, denotes the space of sequences $x^*_m = (m, x_0, \ldots, x_n, \ldots, x_m)$ whose initial segment $(n, x_0, \ldots, x_n) = x^*$ is fixed. We define the σ-algebras $\mathscr{B}^*_{m,x^*}$ correspondingly, and set

$$\mathscr{B}^*_{x^*} = \sigma\Big(\bigcup_{m \geqq |x^*|} \mathscr{B}^*_{m,x^*}\Big).$$

Further, let

$$\mathscr{F}^*_m = \sigma\{\omega : x^*_0, \ldots, x^*_m\} = \sigma\{\omega : x_0, \ldots, x_m\} \in \mathscr{F}_m$$

and suppose $\mathscr{F}^*_{m,x^*}$, $|x^*| = n \leqq m$, is the σ-algebra generated by the values $x^*_{n+1}, \ldots, x^*_m$ for which the initial segments $(n, x_0, \ldots, x_n) = x^*$ are fixed.

Let $\mathbf{P}^*_{x^*}$ be the measure on the sets

$$\mathscr{F}^*_{x^*} = \sigma\Big(\bigcup_{m \geqq |x^*|} \mathscr{F}^*_{m,x^*}\Big),$$

induced in the natural way by the measures $\mathbf{P}_x$, $x \in E$. It can easily be seen that the process $X^* = (x^*_n, \mathscr{F}^*_n, \mathbf{P}^*_x)$, $n \in \boldsymbol{N}$, forms a homogeneous Markov chain in the phase space $(E^*, \mathscr{B}^*)$. For $x^* = (n, x_0, \ldots, x_n)$ we set $g(x^*) = G(n, x_0, \ldots, x_n)$ and let

$$s^*(x^*) = \sup \mathbf{E}^*_{x^*} G(x^*_\tau),$$

where the upper bound is taken over those $\tau \in \mathfrak{R}^{(|x^*|)}_{g,c} \subseteq \mathfrak{R}_{g,c}$ (with respect to $F = \{\mathscr{F}_n\} = \{\mathscr{F}^*_n\}$, $n \in \boldsymbol{N}$) for which $\tau \geqq |x^*|$, and $\mathbf{E}^*_{x^*}$ denotes the mean with respect to the measure $\mathbf{P}^*_{x^*}$.

If $\mathbf{E}_x[\sup_n G^-(n, x_0, \ldots, x_n)] < \infty$, $x \in E$, then, using (2.110), we can easily establish that for all n

$$s^*(n, x_0, \ldots, x_n) = \alpha^n s(x_n) - \sum_{s=0}^{n-1} \alpha^s c(x_s), \tag{2.117}$$

where $s(x) = s^*(0, x)$.

On the basis of Remark 1 of Theorem 3 we can easily show that (2.117) is also valid without the assumption $\mathbf{E}_x[\sup_n G^-(n, x_0, \ldots, x_n)] < \infty$, $x \in E$. Since, according to Theorem 3, $s^*(x^*)$ is the smallest regular excessive majorant of the function $G(x^*)$, it immediately follows that $s(x) = s^*(0, x)$ is the smallest regular (α, c)-excessive majorant of the function $g(x) = G(0, x)$.

Comparing (2.117) with (2.111), we conclude that under the assumption $\mathbf{E}_x[\sup_n G^+(n, x_0, \ldots, x_n)] < \infty$, $x \in E$, the time

$$\begin{aligned}\tau_\varepsilon = \inf\{n \geqq 0 : s^*(n, x_0, \ldots, x_n) \leqq G^*(n, x_0, \ldots, x_n) + \varepsilon\} \\ = \inf\{n \geqq 0 : \alpha^n s(x_n) \leqq \alpha^n g(x_n) + \varepsilon\}\end{aligned}$$

is an ε-optimal stopping time for all $\varepsilon > 0$. From Theorem 3 it also follows that if

τ_0 is a stopping time, then this time is optimal. Fomula (2.116) can be derived easily from (2.110) and (2.117).

COROLLARY. *Suppose* $g(x) \geqq 0$, $\mathbf{E}_x[\sup_n g(x_n)] < \infty$, $c(x) \geqq 0$ *and with* $\mathbf{P}_x$-*probability one*

$$\lim_{n\to\infty} \sum_{s=0}^{n-1} \alpha^s c(x_s) = \infty. \tag{2.118}$$

Then $\tau_0 = \inf\{n\colon s(x_n) = g(x_n)\}$ *is an optimal stopping time.*

For proof it suffices to use the Corollary to Theorem 4.

REMARK 1. Suppose $g \in L$, $c \in L$ and

$$\mathbf{E}_x \sum_{s=0}^{\infty} \alpha^s \,|\, c(x_s) \,|\, < \infty, \qquad x \in E. \tag{2.119}$$

Then

$$s(x) = \sup_{\tau\in\mathfrak{N}} \mathbf{E}_x\left\{\alpha^\tau g(x_\tau) - \sum_{s=0}^{\tau-1} \alpha^s c(x_s)\right\} = \sup_{\tau\in\mathfrak{N}} \mathbf{E}_x \alpha^\tau G(x_\tau) - f(x), \tag{2.120}$$

where

$$f(x) = \mathbf{E}_x \sum_{s=0}^{\infty} \alpha^s c(x_s), \qquad G(x) = g(x) + f(x).$$

Thus, under the assumption (2.119), the problem with fee $c(x) \neq 0$ can be reduced to the solution of a new problem with fee $c(x) \equiv 0$.

To prove the representation (2.120) we let $\xi(\omega) = \sum_{n=0}^{\infty} \alpha^n c(x_n(\omega))$. Then for every $\tau \in \mathfrak{N}$

$$\theta_\tau \xi(\omega) = \sum_{n=0}^{\infty} \alpha^n c(x_{n+\tau}) = \alpha^\tau \sum_{n=\tau}^{\infty} \alpha^n c(x_n)$$

and, by the strong Markov property

$$\begin{aligned} f(x) = \mathbf{E}_x \xi(\omega) &= \mathbf{E}_x \sum_{n=0}^{\infty} \alpha^n c(x_n(\omega)) \\ &= \mathbf{E}_x\left\{\sum_{n=0}^{\tau-1} \alpha^n c(x_n) + \sum_{n=\tau}^{\infty} \alpha^n c(x_n)\right\} \\ &= \mathbf{E}_x \sum_{n=0}^{\tau-1} \alpha^n c(x_n) + \mathbf{E}_x \alpha^\tau \theta_\tau \xi(\omega) \\ &= \mathbf{E}_x \sum_{n=0}^{\tau-1} \alpha^n c(x_n) + \mathbf{E}_x \alpha^\tau \mathbf{E}_{x_\tau} \xi(\omega) \\ &= \mathbf{E}_x \sum_{n=0}^{\tau-1} \alpha^n c(x_n) + \mathbf{E}_x \alpha^\tau f(x_\tau), \end{aligned}$$

from which (2.120) follows.

REMARK 2. The following method of reducing problems with fee $c(x) \neq 0$ to problems in which the fee $c(x) \equiv 0$ may turn out to be useful in those cases in which condition (2.119) is not satisfied.

Let $f = f(x)$ be a solution of the equation

$$\alpha Tf(x) - f(x) = c(x),$$

where $0 < \alpha \leqq 1$, $c(x) \in L$ and $f(x) \in L$.

As in the proof of Theorem 1.5, it can be established that the equality

$$f(x) = \mathbf{E}_x\left\{\alpha^\tau f(x_\tau) - \sum_{s=0}^{\tau-1} \alpha^s c(x_s)\right\}$$

holds for every Markov time $\tau = \tau(\omega)$ such that $\mathbf{P}_x(\tau \leqq N) = 1$, $x \in E$, $N < \infty$.

Suppose $\tau \in \mathfrak{M}$ $\big(\mathbf{P}_x(\tau < \infty) = 1,\ x \in E\big)$. Then, letting $\tau_N = \min(\tau, N)$, $N \in \boldsymbol{N}$, we obtain

$$f(x) = \int_{(\tau \leqq N)} \left[\alpha^\tau f(x_\tau) - \sum_{s=0}^{\tau-1} \alpha^s c(x_s)\right] d\mathbf{P}_x + \int_{(\tau > N)} \left[\alpha^N f(x_N) - \sum_{s=0}^{N-1} \alpha^s c(x_s)\right] d\mathbf{P}_x.$$

If we now assume that the following conditions are satisfied $\big($cf. (1.37) and (1.38)$\big)$

$$\mathbf{E}_x \sum_{s=0}^{\tau-1} \alpha^s \,|\, c(x_s) \,|\, < \infty, \quad \mathbf{E}_x \alpha^\tau \,|\, f(x_\tau) \,|\, < \infty, \qquad (2.121)$$
$$\lim_{N \to \infty} \int_{(\tau > N)} \alpha^N f(x^N) d\mathbf{P}_x = 0,$$

then, passing to the limit ($N \to \infty$) in the preceding equality, we obtain

$$f(x) = \mathbf{E}_x\left\{\alpha^\tau f(x_\tau) - \sum_{s=0}^{\tau-1} \alpha^s c(x_s)\right\}. \qquad (2.122)$$

Consequently, if $\mathfrak{M}^*$ is the class of Markov times $\tau \in \mathfrak{M}$ for which the conditions (2.121) are satisfied, then

$$\begin{aligned} s^*(x) &= \sup_{\tau \in \mathfrak{M}^*} \mathbf{E}_x\left\{\alpha^\tau g(x_\tau) - \sum_{s=0}^{\tau-1} \alpha^s c(x_s)\right\} \\ &= f(x) + \sup_{\tau \in \mathfrak{M}^*} \mathbf{E}_x\{\alpha^\tau G(x_\tau)\}, \end{aligned}$$

where $G(x) = g(x) - f(x)$.

In particular, if $|\, f(x) \,| \leqq K < \infty$ and $|\, c(x) \,| \leqq c < \infty$, then the conditions (2.121) are automatically satisfied for all Markov times τ for which $\mathbf{E}_x \tau < \infty$, $x \in E$.

In conclusion, we consider one example.

EXAMPLE. Suppose $\xi, \xi_1, \xi_2, \cdots$ is a sequence of independent identically distributed random variables with distribution function $F(x)$, and let $\mathbf{E}|\xi| < \infty$. For $x \in E = R$ let $x_n = \max(x, \xi_1, \dots, \xi_n)$, $x_0 = x$, and

$$s(x) = \sup \mathbf{E}_x \left[\alpha^\tau x_\tau - c \sum_{s=0}^{\tau-1} \alpha^s \right],$$

where c is a nonnegative constant and $0 < \alpha \leqq 1$.

It is obvious that $X = (x_n, \mathscr{F}_n, \mathbf{P}_x)$, where $\mathscr{F}_n = \sigma\{\omega: \xi_1, \dots, \xi_n\}$ and $\mathbf{P}_x$ is a measure on sets from $\mathscr{F} = \sigma(\bigcup_n \mathscr{F}_n)$ (induced in the natural fashion by the distributions of the random variables $\xi_1, \xi_2, \cdots$), is a Markov process.

We let γ denote the (unique) root of the equation

$$\mathbf{E}(\xi - \gamma)^+ = \frac{(1 - \alpha)\gamma + c}{\alpha},$$

where

$$\mathbf{E}(\xi - \gamma)^+ = \int_{-\infty}^{\infty} (x - \gamma)^+ \, dF(x) = \int_{(x \geqq \gamma)} (x - \gamma) dF(x),$$

and show that in the class $\mathfrak{M}_N$ the optimal time

$$\tau_N = \min\{n \leqq N: x_n \geqq \gamma\} \tag{2.123}$$

($\tau_N = N$ if the set on the right side of (2.123) is empty), where it is essential that the threshhold γ does not depend on the time $n \leqq N$.

Suppose

$$s_N(x) = \sup_{\tau \in \mathfrak{M}_N} \mathbf{E}_x \left[\alpha^\tau x_\tau - c \sum_{s=1}^{\tau-1} \alpha^s \right] \quad \text{and} \quad Q_{(\alpha,c)} g(x) = \max\{g(x), \alpha T g(x) - c\}.$$

It is clear that $s_N(x) = Q^N_{(\alpha,c)} g(x)$, where $g(x) = x$.

In our case

$$\begin{aligned} s_1(x) = Q_{(\alpha,c)} g(x) &= \max\{x, \alpha \mathbf{E} \max(\xi, x) - c\} \\ &= x + \max\{0, \alpha \mathbf{E}[\max(\xi, x) - x] - (1 - \alpha)x - c\} \\ &= x + \max\{0, \alpha \mathbf{E}(\xi - x)^+ - (1 - \alpha)x - c\}, \end{aligned}$$

whence it can be seen that $s_1(x) = x$ if $x \geqq \gamma$, and $s_1(x) > x$ if $x < \gamma$. This proves the optimality of τ_1.

Analogously, for $s_2(x) = Q^2_{(\alpha,c)} g(x)$ we obtain

$$\begin{aligned} s_2(x) &= \max\{s_1(x), \alpha \mathbf{E} s_1(\max(x, \xi)) - c\} \\ &= s_1(x) + \max\{0, \alpha \mathbf{E}[s_1(\max(x, \xi)) - s_1(x)] - (1 - \alpha) s_1(x) - c\}. \end{aligned}$$

We shall show that $s_2(x) = x$ for all $x \geqq \gamma$ and $s_2 > x$, $x < \gamma$. Indeed, suppose γ_2 is the minimal root of the equation $s_2(x) = x$. It is obvious that $s_2(x) \geqq s_1(x) \geqq x$; therefore $\gamma_2 \geqq \gamma$. But at the point $x = \gamma$

$$\alpha \mathbf{E}[s_1(\max(\gamma,\xi)) - s_1(\gamma)] - (1-\alpha)s_1(\gamma) - c$$
$$= \alpha \mathbf{E}[\max(\gamma,\xi) - \gamma] - (1-\alpha)\gamma - c = 0,$$

and therefore $\gamma_2 = \gamma$.

By induction we can further establish that for all N

$$s_N(x) = x,\ x \geqq \gamma, \quad \text{and} \quad s_N(x) > x,\ x < \gamma,$$

whence follows the optimality of the time τ_N for all $0 < \alpha \leqq 1$ and $N \in \boldsymbol{N}$.

If $0 < \alpha < 1$, then by (2.116) $s_N(x) \uparrow s(x)$ as $N \to \infty$, and the set

$$\{x: s(x) = x\} = \{x: s_N(x) = x\} = \{x: x \geqq \gamma\}.$$

Assuming in addition that $\mathbf{E}_x[\sup_n \alpha^n \max(\xi_1,\dots,\xi_n)] < \infty$ and that $\tau_0 = \inf\{n: x_n \geqq \gamma\}$ belongs to the class $\mathfrak{N}$, we obtain from Theorem 16 that this time is an optimal stopping time.

For example, if $\mathbf{P}\{\xi \leqq a\} = 1$, $a < \infty$, and $\mathbf{P}\{\xi > \gamma\} > 0$, then the time τ_0 is an optimal s.t.

If $\alpha = 1$ and $\mathbf{E}_x[\sup_n \alpha^n \max(\xi_1,\dots,\xi_n)] < \infty$, then from the Corollary to Theorem 16 we conclude that $\tau_0 = \inf\{n: s(x_n) = x_n\}$ is an optimal stopping time. Using Remark 1 to Theorem 3, we can show that $\{x: s(x) = x\} = \{x: x \geqq \gamma\}$. Therefore $\tau_0 = \inf\{n: x_n \geqq \gamma\}$.

CHAPTER III

OPTIMAL STOPPING OF MARKOV RANDOM PROCESSES

§ 1. Excessive functions and their properties

1. As in the case of discrete time, it is natural to believe that "value" also allows an "excessive" characterization for a wide class of Markov processes (with continuous time). This is indeed so; however, the establishment of this fact, and also the investigation of problems of the existence and structure of ε-optimal times, requires the use of rather subtle results from the theory of Markov processes concerning properties of excessive functions.

During this entire chapter we assume that the homogeneous Markov processes X under consideration are *standard* (see § 3 of Chapter I). For simplicity of presentation we also assume that the process X is nonterminating. The changes in statements and proofs in the case of terminating processes (as in the case of discrete time) do not cause essential difficulties.

2. Before proceeding with the "excessive" characterization of value, we introduce the necessary definitions and investigate the properties of excessive functions for the case of continuous time $t \in \boldsymbol{T}$.

Suppose $X = (x_t, \mathscr{F}_t, \mathbf{P}_x)$, $t \in \boldsymbol{T}$, is a (homogeneous, nonterminating, standard) Markov process and $\{T_t\}$, $t \in \boldsymbol{T}$, is the semigroup of operators corresponding to the process X.

DEFINITION 1. The $\overline{\mathscr{B}}$-measurable function $f = f(x)$ such that

$$-\infty < f(x) \leqq \infty \quad \text{and} \quad \mathbf{E}_x f^-(x_t) < \infty, \quad t \in \boldsymbol{T}, \quad x \in E,$$

is said to be *excessive* (for the semigroup $\{T_t\}$, $t \in \boldsymbol{T}$) if

$$T_t f(x) \leqq f(x) \qquad \text{for all } t \in \boldsymbol{T}, \quad x \in E, \tag{3.1}$$

$$\lim_{t \downarrow 0} T_t f(x) = f(x) \qquad \text{for all } x \in E. \tag{3.2}$$

We note the basic properties of the set $\mathscr{E}$ of excessive functions (e.f.) for standard processes. We let $\mathscr{E}^+ \subseteq \mathscr{E}$ denote the set of nonnegative e.f.'s.

I. The function $f(x) \equiv c = \text{const}$ is excessive.[1)]

[1)] For terminating processes, property (3.2) follows from the condition $\lim_{t \downarrow 0} P(t,x,E) = 1$ involved in the definition of a standard process.

II. If $f, g \in \mathscr{E}$ and the constants $a, b \geqq 0$, then $af + bg \in \mathscr{E}$.

III. If $f \in \mathscr{E}$, then $T_t f(x) = \mathbf{E}_x f(x_t) \in \mathscr{E}$, $t \in \boldsymbol{T}$, where $T_t f(x) \geqq T_s f(x)$, $t \leqq s$.

IV. If $f_n \in \mathscr{E}$, $n = 1,2,\cdots$, and $f_{n+1} \geqq f_n$, then the function $f(x) = \lim_{n\to\infty} f_n(x) \in \mathscr{E}$.

In order to formulate the next important property of excessive functions, which shows in particular that the requirement of $\bar{\mathscr{B}}$-measurability of e.f.'s (for standard processes) can be weakened, we introduce a few new concepts.

Suppose μ is a probability measure on $(E, \mathscr{B})$ and

$$\mathbf{P}_\mu(A) = \int_E \mathbf{P}_x(A)\, \mu(dx), \qquad A \in \mathscr{F}.$$

Let $\Gamma \in \hat{\mathscr{B}}$ if for all probability measures μ on $(E,\mathscr{B})$ there are (Borel) sets Γ_1 and Γ_2 from $\mathscr{B}$ such that $\Gamma_1 \subseteq \Gamma \subseteq \Gamma_2$ and

$$\mathbf{P}_\mu\{\chi_{\Gamma_1}(x_t) = \chi_{\Gamma_2}(x_t) \quad \text{for all } t \in \boldsymbol{T}\} = 1,$$

where $\chi_A(x)$ is the indicator function of the set A.

The system $\hat{\mathscr{B}}$ forms a σ-algebra, $\mathscr{B} \subseteq \hat{\mathscr{B}} \subseteq \bar{\mathscr{B}}$; sets from $\hat{\mathscr{B}}$ are said to be *almost Borel*. $\hat{\mathscr{B}}$-measurable functions are also called almost Borel.

V. If $f \in \mathscr{E}$, then the system $(f(x_t), \mathscr{F}_t, \mathbf{P}_x)$, $t \in \boldsymbol{T}$, forms a generalized supermartingale for every $x \in E$, i.e.

$$\mathbf{E}_x f^-(x_t) < \infty, \quad \mathbf{E}_x[f(x_t) \mid \mathscr{F}_s] \leqq f(x), \qquad s \leqq t, \qquad (\mathbf{P}_x\text{-a.s.}).$$

Suppose $X = (x_t, \mathscr{F}_t, \mathbf{P}_x)$, $t \in \boldsymbol{T}$, is a standard process, $f \in \mathscr{E}$ and there exists an $\mathscr{F}$-measurable integrable random variable $\eta = \eta(\omega)$ such that

$$f(x_t) \geqq \mathbf{E}_x[\eta \mid \mathscr{F}_t], \qquad t \in \boldsymbol{T}, \quad x \in E. \tag{3.3}$$

Then the function $f(x)$ is almost Borel, the process $f(x_t(\omega))$ is continuous[2)] from the right ($\mathbf{P}_x$-a.s., $x \in E$), and for all $t \in (0,\infty)$ the limit $\lim_{u\uparrow t} f(x_u)$ exists for all $x \in E$ with $\mathbf{P}_x$-probability one. (For proofs of property V see, for example, [**40**], § 5, [**25**], Theorems 12.4 and 12.6, [**50**], Chapter XIV, or [**12**], Chapter II.)

According to property V, for standard processes Definition 1 can be replaced (at least under the assumption (3.3)) by its equivalent

DEFINITION 2. An almost Borel function $f = f(x)$ such that

$$-\infty < f(x) \leqq \infty \quad \text{and} \quad \mathbf{E}_x[f^-(x_t)] < \infty, \qquad t \in \boldsymbol{T}, \quad x \in E,$$

is said to be *excessive* if conditions (3.1) and (3.2) are satisfied.

DEFINITION 3. An almost Borel function $f = f(x)$ is said to be $\mathscr{C}$-*lower* (*upper*) *semicontinuous* (or just lower [upper] semicontinuous in the natural topology as-

---

2) Since an excessive function may take on the value $+\infty$, continuity is defined in the topology of the extended number line.

sociated with the process X) if

$$\liminf_{t\downarrow 0} f(x_t) \geqq f(x) \quad \Big(\limsup_{t\downarrow 0} f(x_t) \leqq f(x)\Big) \qquad (\mathbf{P}_x\text{-a.s., } x \in E). \tag{3.4}$$

REMARK. It is known ([25], Theorem 4.9 and the remark after Theorem 12.4) that for standard processes a $\mathscr{C}_0$-lower semicontinuous excessive function $f(x)$ is indeed $\mathscr{C}_0$-continuous, i.e.

$$\lim_{t\downarrow 0} f(x_t) = f(x) \qquad (\mathbf{P}_x\text{-a.s., } x \in E).$$

It follows from property V that for standard processes satisfying (3.3), Definitions 1 and 2 are equivalent to the following.

DEFINITION 4. An almost Borel function $f = f(x)$ such that

$$-\infty < f(x) \leqq \infty \quad \text{and} \quad \mathbf{E}_x f^-(x_t) < \infty, \qquad t \in \boldsymbol{T}, \quad x \in E,$$

is said to be *excessive* (for the process X with semigroup $\{T_t\}$, $t \in \boldsymbol{T}$) if $T_t f(x) \leqq f(x)$ $(t \in \boldsymbol{T},\ x \in E)$ and

$$f(x) \text{ is a } \mathscr{C}_0\text{-continuous function.} \tag{3.5}$$

In view of the remarks made above, it suffices to require $\mathscr{C}_0$-lower semicontinuity, instead of $\mathscr{C}_0$-continuity, for $f(x)$.

VI. Suppose X is a standard process. If $f, g \in \mathscr{E}^+$, then the function $f \wedge g = \min(f, g) \in \mathscr{E}^+$. If $f \in \mathscr{E}$ and c is a constant, then $f^c = \min(f, c) \in \mathscr{E}$.

VII. If $f \in \mathscr{E}$ and $\sup_{t\in \boldsymbol{T}} \mathbf{E}_x f^-(x_t) < \infty$, $x \in E$, then with $\mathbf{P}_x$-probability one the limit $\lim_{t\to\infty} f(x_t(\omega))$, finite or equal to $+\infty$, exists.

This property follows from Theorem 1.3.

§ 2. Smallest excessive majorants and their construction

1. Suppose $X = (x_t, \mathscr{F}_t, \mathbf{P}_x)$, $t \in \boldsymbol{T}$, $x \in E$, is a (homogeneous, standard) Markov process. Let L denote the class of almost Borel $\mathscr{C}_0$-lower semicontinuous functions $g = g(x)$ such that

$$-\infty < g(x) \leqq \infty, \quad \mathbf{E}_x g^-(x_t) < \infty, \qquad t \in \boldsymbol{T}, \quad x \in E.$$

We note that if $g \in L$ is $\mathscr{C}_0$-continuous, then according to Theorem 4.11 of **[25]**, the process $g(x_t)$, $t \in \boldsymbol{T}$, has $\mathbf{P}_x$-a.s., $x \in E$, right-continuous trajectories.

We let $L(A^-)$ and $L(A^+)$ denote the classes of functions $g = g(x)$ from L for which the process $g(x_t)$, $t \in \boldsymbol{T}$, is separable[3] (**[22]**,**[25]**) and the following conditions are satisfied respectively:

[3] The condition of separability (it is satisfied if, for example, $\mathbf{P}_x$-a.s. the trajectories of the process $g(x_t)$, $t \in \boldsymbol{T}$, are right continuous, $x \in E$) guarantees the $\mathscr{F}$-measurability of the variables $\sup_{t\in \boldsymbol{T}} g(x_t)$ and $\inf_{t\in \boldsymbol{T}} g(x_t)$.

$$A^-: \mathbf{E}_x\Big[\sup_{t\in\boldsymbol{T}} g^-(x_t)\Big] < \infty, \qquad x \in E, \tag{3.6}$$
$$A^+: \mathbf{E}_x\Big[\sup_{t\in\boldsymbol{T}} g^+(x_t)\Big] < \infty, \qquad x \in E.$$

We also set $L(A^-,A^+) = L(A^-) \cap L(A^+)$.

Let $\overline{\mathfrak{M}} = \{\tau\}$ be the class of all Markov times (M.t.'s) (relative to $F = \{\mathscr{F}_t\}$, $t \in \boldsymbol{T}$), and let $\mathfrak{M} \subseteq \overline{\mathfrak{M}}$ be the class of finite Markov times τ $(\mathbf{P}_x(\tau < \infty) = 1$, $x \in E)$, which we also call *stopping times* (s.t.'s).

As in the case of discrete time, for functions $g \in L(A^-)$ we define the values

$$s(x) = \sup_{\tau\in\mathfrak{M}} \mathbf{E}_x[g(x_\tau); \tau < \infty], \tag{3.7}$$
$$\bar{s}(x) = \sup_{\tau\in\overline{\mathfrak{M}}} \mathbf{E}_x[g(x_\tau); \tau < \infty], \qquad g(x) \geqq 0,$$
$$\tilde{s}(x) = \sup_{\tau\in\overline{\mathfrak{M}}} \tilde{\mathbf{E}}_x g(x_\tau),$$

where

$$\tilde{\mathbf{E}}_x g(x_\tau) = \mathbf{E}_x[g(x_\tau); \tau < \infty] + \mathbf{E}_x\Big[\limsup_t g(x_t); \tau = \infty\Big]. \tag{3.8}$$

We define (ε,s)-, $(\varepsilon,\bar{s})$- and $(\varepsilon,\tilde{s})$-optimal Markov times analogously to §2 of Chapter II.

DEFINITION 1. Suppose $g \in L(A^-)$. The excessive function $f \in L$ is said to be an *excessive majorant* (e.m.) of $g(x)$ if $f(x) \geqq g(x)$, $x \in E$. The function $f(x)$ is said to be the *smallest* e.m. (s.e.m.) of $g(x)$ if $f(x)$ is an e.m. and $f(x) \leqq h(x)$, where $h(x)$ is an arbitrary e.m. of $g(x)$.

In order to justify the last definition, we shall show that the s.e.m. of the arbitrary function $g \in L(A^-)$ indeed exists.

Suppose $g \in L(A^-)$. We let

$$Q_n g(x) = \max\{g(x), T_{2^{-n}} g(x)\}, \qquad n \in \boldsymbol{N}, \tag{3.9}$$

and

$$v(x) = \lim_{n\to\infty} \lim_{N\to\infty} Q_n^N g(x), \tag{3.10}$$

where Q_n^N is the Nth power of the operator Q_n, $n \in \boldsymbol{N}$.

LEMMA 1. *If $g \in L(A^-)$, then the function $v(x)$ defined by formula* (3.10) *is the s.e.m. of $g(x)$.*

PROOF. Let $v_n(x) = \lim_{N\to\infty} Q_n^N g(x)$. By Theorem II. 1 and Lemma II. 4,

$$v_n(x) = \sup_{\tau\in\mathfrak{M}(n)} \mathbf{E}_x g(x_\tau), \tag{3.11}$$

where $\mathfrak{M}(n)$ is the class of stopping times taking on the values $k \cdot 2^{-n}$, $k \in \boldsymbol{N}$, and such that

$$\{\tau = k \cdot 2^{-n}\} \in \mathscr{F}^{(n)}_{k \cdot 2^{-n}} = \sigma\{\omega : x_0, x_{2^{-n}}, \ldots, x_{k \cdot 2^{-n}}\}.$$

Since $\mathfrak{M}(n+1) \supseteq \mathfrak{M}(n)$, it follows that $v_{n+1}(x) \geqq v_n(x)$, and consequently the limit $\lim_{n\to\infty} v_n(x)$, which we denote by $v(x)$ in (3.10), exists.

It is clear that $v(x) \geqq g(x)$, $v_n(x) \geqq T_{2^{-n}}v_n(x)$ and for all $m \in \boldsymbol{N}$

$$v_n(x) \geqq T_{m \cdot 2^{-n}} v_n(x).$$

We take $m = l \cdot 2^{n-k}$, $l \in \boldsymbol{N}$. Then $v_n(x) \geqq T_{l \cdot 2^{-k}} v_n(x)$ and

$$v(x) \geqq T_{l \cdot 2^{-k}} v(x). \tag{3.12}$$

We shall show that $v(x)$ is $\mathscr{C}_0$-lower semicontinuous. To this end we consider an arbitrary function $\varphi(x) \in L(A^-)$. Let $\Phi(x) = T_t\varphi(x)$, where $t \in \boldsymbol{T}$.

Suppose τ_n are the times of first entry of some compacta, and let $\mathbf{P}_x(\tau_n \downarrow 0) = 1$. Then, by (1.21),

$$\mathbf{E}_x\Phi(x_{\tau_n}) = \mathbf{E}_x\mathbf{E}_{x_{\tau_n}}\varphi(x_t) = \mathbf{E}_x\theta_{\tau_n}\varphi(x_t) = \mathbf{E}_x\varphi(x_{\tau_n+t})$$

and by Fatou's Lemma,

$$\begin{aligned} \liminf_{n\to\infty} \mathbf{E}_x\Phi(x_{\tau_n}) &= \liminf_{n\to\infty} \mathbf{E}_x\varphi(x_{\tau_n+t}) \\ &\geqq \mathbf{E}_x \liminf_{n\to\infty} \varphi(x_{\tau_n+t}) \geqq \mathbf{E}_x\varphi(x_t) = \Phi(x). \end{aligned} \tag{3.13}$$

Since the function $\varphi(x)$ is almost Borel, from this property it is easy to derive that the function

$$\Phi(x) = \mathbf{E}_x\varphi(x_t) = \int_\Omega \varphi(x_t(\omega))\mathbf{P}_x(d(\omega))$$

is also almost Borel. But it is known (see Theorem 4.9 in **[25]**) that an almost Borel function $\Phi(x)$ satisfying (3.13) is $\mathscr{C}_0$-lower semicontinuous.

Thus the (almost Borel) functions

$$\begin{gathered} T_{2^{-n}}g(x), \qquad Q_n g(x) = \max\{g(x), T_{2^{-n}}g(x)\}, \\ v_n(x) = \lim_{N\to\infty} Q_n^N g(x), \qquad v(x) = \lim_{n\to\infty} v_n(x) \end{gathered}$$

are $\mathscr{C}_0$-lower semicontinuous.

We establish the inequality $v(x) \geqq T_t v(x)$, $t \geqq 0$. We take a sequence of binary rational numbers $r_i \downarrow t$ as $i \to \infty$. Using in succession (3.12), the right continuity of the trajectories of the process X, the $\mathscr{C}_0$-lower semicontinuity of $v(x)$, the membership of the function $v(x)$ in the class $L(A^-)$ and finally Fatou's Lemma, we obtain

$$\begin{aligned} v(x) &\geqq \liminf_{i\to\infty} T_{r_i} v(x) \\ &= \liminf_{i\to\infty} \mathbf{E}_x v(x_{r_i}) \geqq \mathbf{E}_x \liminf_{i\to\infty} v(x_{r_i}) \geqq \mathbf{E}_x v(x_t) = T_t v(x), \end{aligned}$$

which proves the excessiveness of $v(x)$.

Now suppose that $u(x)$ is an e.m. of $g(x)$. Then it follows from the inequality $u(x) \geqq g(x)$ that $u(x) = Q_n^N(x) \geqq Q_n^N g(x)$. Therefore $u(x) \geqq v(x)$, and consequently $v(x)$ is the s.e.m. of the function $g(x)$. The lemma has been proved.

REMARK 1. Let

$$g \in L(A^-), \quad g^b(x) = \min(b, g(x)), \quad b \geqq 0.$$

Then for the s.e.m. $v(x)$ of the function $g(x)$ we have

$$v(x) = \lim_{n\to\infty} \lim_{b\to\infty} \lim_{N\to\infty} Q_n^N g^b(x) = \lim_{n\to\infty} \lim_{N\to\infty} \lim_{b\to\infty} Q_n^N g^b(x).$$

The proof of these equalities follows from Remark 2 to Lemma I.4 and formula (3.10).

2. The lemmas proved in this section yield additional information on the structure of the s.e.m. of the function $g(x)$ in the case where the process X is a *Feller* process.

LEMMA 2. *Suppose X is a standard Feller process, and the function $g(x) \geqq C > -\infty$ is continuous. Then its s.e.m. $v(x)$ is a function which is lower semicontinuous* $(\liminf_{y\to x} v(y) \geqq v(x))$.

PROOF. We can assume without loss of generality that the function $g(x)$ is nonnegative. Since the function $g(x)$ is continuous, each of the bounded functions $g^m(x) = \min(m, g(x))$, $m \in \boldsymbol{N}$, is also continuous. Since X is a Feller process, the functions $T_t g^m(x)$, $t \in \boldsymbol{T}$, are continuous. Hence (see the proof of the preceding lemma) it follows that each of the functions $Q_n g^m(x)$, $Q_n^N g^m(x)$ is continuous. Therefore the functions

$$v_n^m(x) = \lim_{N\to\infty} Q_n^N g^m(x) \quad \text{and} \quad v^m(x) = \lim_{n\to\infty} v_n^m(x)$$

are lower semicontinuous (as the limit of a monotone increasing sequence of continuous functions).

Since $v^{m+1}(x) \geqq v^m(x)$, the function $\bar{v}(x) = \lim_{m\to\infty} v^m(x)$ is also lower semicontinuous (as the limit of a monotone increasing sequence of lower semicontinuous functions).

It remains to show that $\bar{v}(x) = v(x)$. But this equality can be established in precisely the same way that the analogous relationship was proved in Theorem II. 1.

REMARK. Lemma 2 also holds for continuous functions $g \in L(A^-)$ as long as the functions $T_t g^m(x)$ are continuous for all $t \in \boldsymbol{T}$ and $m \in \boldsymbol{N}$.

To find the s.e.m. of nonnegative continuous functions $g(x)$, the following method of construction often turns out to be of value.

Suppose (cf. (2.12))

$$\tilde{Q}g(x) = \sup_{t \geqq 0} T_t g(x), \qquad \tilde{Q}^0 g(x) = g(x) \tag{3.14}$$

and

$$\tilde{Q}^N g(x) = \sup_{t \geqq 0} T_t(\tilde{Q}^{N-1}g)(x),$$

where $\tilde{Q}^N$ is the Nth power of $\tilde{Q}$.

LEMMA 3. *Suppose X is a standard Feller process, and the function $g(x) \geqq C > -\infty$ is continuous. Then the function*

$$\upsilon(x) = \lim_{N \to \infty} \tilde{Q}^N g(x) \tag{3.15}$$

is lower semicontinuous and is the s.e.m. of the function $g(x)$.

PROOF. Let $\upsilon_N(x) = \tilde{Q}^N g(x)$. Then

$$\upsilon_{N+1}(x) = \tilde{Q}\upsilon_N(x) = \sup_{t \geqq 0} T_t \upsilon_N(x) \geqq \upsilon_N(x) \geqq g(x),$$

and

$$\upsilon_{N+1}(x) \geqq T_t \upsilon_N(x)$$

for all $t \geqq 0$. Since $\upsilon_N(x) \uparrow \upsilon(x)$ as $N \to \infty$, it follows that

$$\upsilon(x) \geqq T_t \upsilon(x), \qquad t \geqq 0, \tag{3.16}$$

and $\upsilon(x) \geqq g(x)$.

We shall show that the function $\upsilon(x)$ is lower semicontinuous. Since $g(x)$ is continuous, the function $g^m(x) = \min(m, g(x))$, $m \in \boldsymbol{N}$, is also continuous. Since X is a Feller process, the function $T_t g^m(x)$ is continuous for all $t \in \boldsymbol{T}$ and $m \in \boldsymbol{N}$. Hence (as in Lemma 2) it follows that the functions $T_t g(x)$, $t \in \boldsymbol{T}$, and $\upsilon_1(x) = \tilde{Q}g(x) = \sup_{t \geqq 0} T_t g(x)$ are lower semicontinuous. We shall show by induction that each of the functions $\upsilon_N(x)$, $N \in \boldsymbol{N}$, is also lower semicontinuous.

Suppose that for some $N \geqq 1$ the function $\upsilon_N(x)$ is lower semicontinuous. Then we shall establish that $\upsilon_{N+1}(x)$ is also lower semicontinuous. To this end we construct a nondecreasing sequence $\{\upsilon_N^i(x)\}$, $i = 1,2,\cdots$, of bounded continuous functions[4] such that $\upsilon_N^i(x) \uparrow \upsilon_N(x)$ as $i \to \infty$. Then the functions $T_t \upsilon_N^i(x)$ are continuous in x, and from the equalities

---

[4] See, for example, [33], Chapter VII, Theorem 30, or [52], Chapter XV, Theorem 10, for a proof that such a construction is possible.

$$v_{N+1}(x) = \sup_{t \geq 0} T_t v_N(x) = \sup_{t \geq 0} \lim_{i \to \infty} T_t v_N^i(x)$$

it follows that the functions $v_{N+1}(x)$, and thus also $v(x) = \lim_{N\to\infty} v_N(x)$, are lower semicontinuous.

Thus $v(x) \geqq g(x)$ and $v(x) \geqq T_t v(x)$, and obviously if $h(x)$ is an e.m. of $g(x)$, then $v(x) = \lim_{N\to\infty} \tilde{Q}^N g(x) \leqq h(x)$. To complete the proof it now remains to establish that $\lim_{t\downarrow 0} T_t v(x) = v(x)$. From (3.16), $v(x) \geqq \limsup_{t\downarrow 0} T_t v(x)$. On the other hand, since the function $v(x)$ is lower semicontinuous and the process X has right continuous ($\mathbf{P}_x$-a.s., $x \in E$) trajectories, Fatou's Lemma implies that

$$\liminf_{t\downarrow 0} T_t v(x) = \liminf_{t\downarrow 0} \mathbf{E}_x v(x_t) \geqq \mathbf{E}_x \liminf_{t\downarrow 0} v(x\,) \geqq v(x).$$

The lemma is completely proved.

3. In the case where the function $g \in L(A^-,A^+)$, we can propose the following method for finding its s.e.m. function.

Let

$$\varphi(x) = \mathbf{E}_x \left[\sup_{t \in \boldsymbol{T}} g(x_t)\right], \qquad \varphi_n(x) = \mathbf{E}_x \left[\sup_{k \in \boldsymbol{N}} g(x_{k\cdot 2^{-n}})\right].$$

If $f \in L$, then we set (cf. (2.19))

$$G_n f(x) = \max\{g(x), T_{2^{-n}} f(x)\}, \qquad n \in \boldsymbol{N}, \tag{3.17}$$

and we let G_n^N be the Nth power of the operator G_n, $G_n^0 f = f$. We note that if $f(x) = g(x)$, then $G_n g(x) = Q_n g(x)$.

LEMMA 4. *If the function* $g \in L(A^-,A^+)$, *then its s.e.m.*

$$v(x) = \lim_{n\to\infty} \lim_{N\to\infty} G_n^N \varphi_n(x). \tag{3.18}$$

PROOF. Let $\bar{v}_n(x) = \lim_{N\to\infty} G_n^N \varphi_n(x)$. According to Lemma II. 9, $\bar{v}_n(x)$ coincides with the function $v_n(x)$ defined in (3.11). Applying Lemma 1, we obtain the desired assertion (3.18).

§ 3. Excessive characterization of value

1. As in the case of discrete time, a fundamental role in the study of the properties of $s(x)$, $\bar{s}(x)$ and $\bar{\bar{s}}(x)$ is played by

LEMMA 5. *Suppose* $X = (x_t, \mathscr{F}_t, \mathbf{P}_x)$, $t \in \boldsymbol{T}$, $x \in E$, *is a standard Markov process and* $f = f(x)$ *is an excessive function satisfying the condition*

$$A^-\colon \mathbf{E}_x \left[\sup_{t \in \boldsymbol{T}} f^-(x_t)\right] < \infty, \qquad x \in E. \tag{3.19}$$

Suppose the Markov times $\tau, \sigma \in \overline{\mathfrak{M}}$, where $\tau \geqq \sigma$ with $\mathbf{P}_x$-probability one for all $x \in E$. Then

$$\tilde{\mathbf{E}}_x f(x_\sigma) \geqq \tilde{\mathbf{E}}_x f(x_\tau), \qquad x \in E, \tag{3.20}$$

and, in particular, for all x in E

$$f(x) \geqq \int_{(\tau<\infty)} f(x_\tau) d\mathbf{P}_x + \int_{(\tau=\infty)} \limsup_t f(x_t) d\mathbf{P}_x. \tag{3.21}$$

PROOF. We note that by property VII the limit $\lim_{t\to\infty} f(x_t)$ exists. Therefore we can write $\lim_t f(x_t)$ instead of $\limsup_t f(x_t)$ in (3.21). Verification of the validity of (3.21) using Theorem I. 5 can be carried out in the same way as in the case of discrete time.

With the aid of this lemma we can prove the following assertion.

LEMMA 6. *Suppose $f = f(x)$ is an excessive function satisfying the condition A^- and*

$$\tau_A = \inf\{t > 0; x_t \in A\}, \tag{3.22}$$

where $A \in \hat{\mathscr{B}}$. Then the function

$$f_A(x) = \mathbf{E}_x f(x_{\tau_A}) \tag{3.23}$$

is also excessive.

Before proceeding with the proof, we note that generally speaking the lemma does not hold for times $\sigma_A = \inf\{t \geqq 0: x_t \in A\}$.

PROOF. Let $s \geqq 0$ and

$$\tau_A^s = \inf\{t > s: x_t \in A\}. \tag{3.24}$$

It follows from Theorem I.2 that the times τ_A^s are Markov times. As in the proof of Lemma II.2, it can be established that $T_t f_A(x) \leqq f_A(x)$, $t \in \boldsymbol{T}$, $x \in E$. From the reasoning used in the proof of Lemma 1 it follows that the function $f_A(x)$ is almost Borel. Therefore we need only verify the relationship

$$\lim_{t\to 0} T_t f_A(x) = f_A(x), \qquad x \in E.$$

Since[5] $\tau_A^s \downarrow \tau_A$ as $s \downarrow 0$, by the right continuity of the process $f_A(x_t)$ and Fatou's Lemma we have

$$\begin{aligned} \liminf_{t\downarrow 0} T_t f_A(x) &= \liminf_{t\downarrow 0} \mathbf{E}_x f(x_{\tau_A^t}) \geqq \mathbf{E}_x \liminf_{t\downarrow 0} (x_{\tau_A^t}) \\ &= \mathbf{E}_x f(x_{\tau_A}) = f_A(x). \end{aligned} \tag{3.25}$$

---

[5] If $\sigma_A^s = \inf\{t \geqq s: x_t \in A\}$, then generally speaking $\sigma_A^s \nrightarrow \sigma_A$ as $s \downarrow 0$.

On the other hand, obviously $\limsup_{t\downarrow 0} T_t f_A(x) \leqq f_A(x)$, which together with (3.25) proves the desired equality.

2. The following result can be proved by the same method as Lemma II.6.

LEMMA 7. *Suppose the function $g \in L$ satisfies the condition*

$$A^+ \colon \mathbf{E}_x \left[\sup_t g^+(x_t)\right] < \infty, \qquad x \in E,$$

and $v(x)$ is its s.e.m. Then

$$\limsup_t v(x_t) = \limsup_t g(x_t). \tag{3.26}$$

3. THEOREM 1. *Suppose $X = (x_t, \mathscr{F}_t, \mathbf{P}_x)$, $t \in \boldsymbol{T}$, is a standard Markov process, and $g(x) \in L(A^-)$. Then the value $s(x)$ is the s.e.m. of the function $g(x)$:*

$$s(x) = \tilde{s}(x) \tag{3.27}$$

and $s(x) = \bar{s}(x) = \tilde{s}(x)$ if $g(x) \geqq 0$.

PROOF. Suppose $v(x)$ is the s.e.m. of the function $g(x) \in L(A^-)$. By Lemma 5, for all $\tau \in \overline{\mathfrak{M}}$

$$\begin{aligned} v(x) &\geqq \mathbf{E}_x[v(x_\tau); \tau < \infty] + \mathbf{E}_x\left[\limsup_t v(x_t); \tau = \infty\right] \\ &\geqq \mathbf{E}_x[g(x_\tau); \tau < \infty] + \mathbf{E}_x\left[\limsup_t g(x_t); \tau = \infty\right], \end{aligned}$$

from which it is obvious that $v(x) \geqq \tilde{s}(x) \geqq s(x)$ and that in the case of nonnegative functions $g(x)$

$$v(x) \geqq \tilde{s}(x) \geqq \bar{s}(x) \geqq s(x).$$

The opposite inequality $v(x) \leqq s(x)$ follows from Lemma 1. Indeed, the class of Markov times $\mathfrak{M}(n) \subseteq \mathfrak{M}$, and therefore (see (3.11)) $v_n(x) \leqq s(x)$. But $v(x) = \lim_{n\to\infty} v_n(x)$, and consequently $v(x) \leqq s(x)$.

COROLLARY. *Suppose the time $\tau^* \in \overline{\mathfrak{M}}$ is such that its corresponding reward $\bar{f}(x) = \mathbf{E}_x g(x_{\tau^*})$ (or $\tilde{f}(x) = \tilde{\mathbf{E}}_x g(x_{\tau^*})$) is an excessive function and $\bar{f}(x) \geqq g(x)$ ($\tilde{f}(x) \geqq g(x)$). Then $\bar{f}(x) = \bar{s}(x)$ ($\tilde{f}(x) = \tilde{s}(x)$) and the time τ^* is $(0,\bar{s})$- (respectively $(0,\tilde{s})$-) optimal. In exactly the same way, if the time $\tau^* \in \mathfrak{M}$, the function $f(x) = \mathbf{E}_x g(x_\tau)$ is excessive and $f(x) \geqq g(x)$, then τ^* is a $(0,s)$-optimal stopping time.*

To illustrate this corollary we consider the following

EXAMPLE. Suppose $W = (w_t, \mathscr{F}_t, \mathbf{P}_x)$, $t \in \boldsymbol{T}$, $x \in R^1$, is a Wiener process (more precisely, a family of Wiener processes with $w_0(\omega) = x$ and $x \in R^1$) such that $\mathbf{P}_x(w_0 = x) = 1$ and

$$\mathbf{E}_x\left[w_{t+\Delta} - w_t\right] = \mu\Delta, \qquad \mathrm{Var}_x\left[w_{t+\Delta} - w_t\right] = \Delta,$$

$$t \geqq 0, \quad \Delta > 0, \quad x \in R^1.$$

We take $g(x) = \max(0,x)$ and let

$$\bar{s}(x) = \sup_{\tau \in \bar{\mathfrak{M}}} \mathbf{E}_x g(x_\tau), \qquad \tilde{s}(x) = \sup_{\tau \in \bar{\mathfrak{M}}} \tilde{\mathbf{E}}_x g(x_\tau).$$

It can easily be seen that for $\mu \geqq 0$ we have $\bar{s}(x) \equiv \infty$ and the time $\bar{\tau}(\omega) \equiv \infty$ is $(0,\bar{s})$-optimal. Now suppose that $\mu < 0$.

Let $\sigma_\gamma = \inf\{t \geqq 0: w_t \in \Gamma_\gamma\}$, $\Gamma_\gamma = [\gamma,\infty)$. As in the example considered in Chapter 2, § 1.10, it can be shown that $\sigma_\gamma \in \bar{\mathfrak{M}}$ and

$$f_\gamma(x) = \mathbf{E}_x g(x_{\sigma_\gamma}) = \begin{cases} \gamma e^{2\mu(\gamma - x)}, & x \leqq \gamma, \\ x, & x \geqq \gamma. \end{cases}$$

Setting $\bar{f}(x) = \sup_{\gamma \in R^1} f_\gamma(x)$, we find that $\bar{f}(x) = f_{\gamma^*}(x)$, where $\gamma^* = -1/2\mu$. It is clear that $f_{\gamma^*}(x) \geqq g(x)$, and direct verification shows that $\bar{f}(x) \geqq T_t\bar{f}(x)$ for all t. Applying the Corollary to Theorem 1, we see that the time $\sigma_{\gamma^*} = \inf\{t \geqq 0: w_t \in \Gamma_{\gamma^*}\}$ is $(0,\bar{s})$-optimal. It is of interest to note that $\mathbf{P}_x(\sigma_{\gamma^*} < \infty) < 1$ for all $x < \gamma^*$, so that the $(0,\bar{s})$-optimal time σ_{γ^*} is not a stopping time.

4. In comparison with the proof of Theorem II.1, the method used here to prove Theorem 1 has the drawback that it does not yield a direct method for constructing ε-optimal stopping times. Since the proof of Theorem II.1 was essentially based on the relationship $v(x) = \mathbf{E}_x v(x_{\tau_\varepsilon})$ (see (2.29)) for the s.e.m. $v(x)$ of the function $g(x) \in L(A^-,A^+)$, it is natural to examine conditions under which it also holds for the case of continuous time. A partial answer to this question is contained in Lemmas 8 and 9 given below.

Let

$$\Gamma_\varepsilon = \{x: v(x) \leqq g(x) + \varepsilon\}, \qquad \varepsilon \geqq 0,$$

where $v(x)$ is the s.e.m. of the (almost Borel) function $g(x)$. Also let

$$\tau_\varepsilon = \inf\{t > 0: x_t \in \Gamma_\varepsilon\}, \qquad \sigma_\varepsilon = \inf\{t \geqq 0: x_t \in \Gamma_\varepsilon\}.$$

Since the functions $v(x)$ and $g(x)$ are $\mathscr{B}$-measurable, the set Γ_ε is almost Borel and, according to Theorem I.2, the times τ_ε and σ_ε are Markov.

LEMMA 8. *Suppose $X = (x_t, \mathscr{F}_t, \mathbf{P}_x)$, $t \in \boldsymbol{T}$, is a standard Markov process, $g(x)$ is a bounded ($|g(x)| \leqq K < \infty$) almost Borel $\mathscr{C}_0$-continuous function and $v(x)$ is its s.e.m. Then for every $\varepsilon > 0$*

$$v(x) = \mathbf{E}_x v(x_{\tau_\varepsilon}) \tag{3.28}$$

and

$$v(x) = \mathbf{E}_x v(x_{\sigma_\varepsilon}). \tag{3.29}$$

PROOF. First we note that, by Lemma 7, $\mathbf{P}_x(\tau_\varepsilon < \infty) = 1$, $x \in E$, for all $\varepsilon > 0$. Therefore τ_ε and σ_ε are stopping times. Further, it suffices to prove only relationship (3.28), since (3.29) follows directly from (3.28) because (see Lemma 5)

$$v(x) \geqq \mathbf{E}_x v(x_{\sigma_\varepsilon}) \geqq \mathbf{E}_x v(x_{\tau_\varepsilon}).$$

Thus we proceed with the proof of (3.28). Let $v_\varepsilon(x) = \mathbf{E}_x v(x_{\tau_\varepsilon})$. According to Lemma 6, $v_\varepsilon(x)$ is an excessive function, and $v_\varepsilon(x) \leqq v(x)$ by Lemma 5. Therefore, when we have proved $v_\varepsilon(x) \geqq g(x)$, we shall immediately obtain the desired equality $v_\varepsilon(x) = v(x)$.

Let

$$c = \sup_{x \in E}[g(x) - v_\varepsilon(x)]. \tag{3.30}$$

We have two cases: $c \leqq 0$ and $c > 0$. In the first case, obviously $v_\varepsilon(x) \geqq g(x)$.

Now suppose

$$0 < c = \sup_{x \in E}[g(x) - v_\varepsilon(x)]. \tag{3.31}$$

We note that $\Gamma_\varepsilon \supseteq \Gamma_{\varepsilon'}$, $\varepsilon \geqq \varepsilon'$, and $\Gamma_0 = \bigcap_{\varepsilon > 0} \Gamma_\varepsilon$. We let $\partial\Gamma_\varepsilon$ denote the boundary of the set Γ_ε. Then for all points $x \in \Gamma_0 \backslash \partial\Gamma_0$, by the right continuity of the process X,

$$v_\varepsilon(x) = \mathbf{E}_x v(x_{\tau_\varepsilon}) = v(x) \geqq g(x).$$

Consequently

$$0 < c = \sup_{x \in E}[g(x) - v_\varepsilon(x)] = \sup_{x \in (E \backslash \Gamma_0) \cup \partial\Gamma_0}[g(x) - v_\varepsilon(x)].$$

The function $c + v_\varepsilon(x)$ is excessive, and $c + v_\varepsilon(x) \geqq g(x)$ for all $x \in E$. Therefore $c + v_\varepsilon(x) \geqq v(x)$. We take $0 < \alpha < \min(c,\varepsilon)$. Since $c < \infty$, there is a point $x_0 \in (E \backslash \Gamma_0) \cup \partial\Gamma_0$ such that

$$g(x_0) - v_\varepsilon(x_0) > c - \alpha, \tag{3.32}$$

whence

$$0 \leqq v(x_0) - g(x_0) \leqq v_\varepsilon(x_0) + c - g(x_0) < \alpha < \varepsilon,$$

i.e.

$$v(x_0) < g(x_0) + \varepsilon. \tag{3.33}$$

Thus the point $x_0 \in \Gamma_\varepsilon \backslash \partial\Gamma_\varepsilon$. But $v_\varepsilon(x_0) = \mathbf{E}_{x_0} v(x_{\tau_\varepsilon}) = v(x_0) \geqq g(x_0)$, which together with (3.32) yields the inequality $\alpha > c$, which contradicts the assertion $0 < \alpha < \min(c,\varepsilon)$. The lemma is proved.

On the basis of this lemma we can now give another proof of the inequality $v(x) \leqq s(x)$ used in the proof of Theorem 1.

Suppose $|g(x)| \leqq K < \infty$. Then, since the functions $g(x)$ and $v(x)$ are $\mathscr{C}_0$-continuous and the process X is right continuous, $x_{\sigma_\varepsilon} \in \Gamma_\varepsilon$. Thus for all $x \in E$

$$s(x) \geqq \mathbf{E}_x g(x_{\sigma_\varepsilon}) \geqq \mathbf{E}_x v(x_{\sigma_\varepsilon}) - \varepsilon = v(x) - \varepsilon. \tag{3.34}$$

Since $\varepsilon > 0$ is arbitrary, we now obtain that $s(x) \geqq v(x)$.

We note that it can be shown, as in the proof of Theorem II.1, that the inequality $s(x) \geqq v(x)$ holds if we require of the function $g(x)$ only that it be bounded from below: $g(x) \geqq K > -\infty$.

COROLLARY 1. *If* $|g(x)| \leqq K < \infty$, *then for each* $\varepsilon > 0$ *the time* σ_ε *is an* (ε,s)-*optimal stopping time.*

Indeed, according to Theorem 1, $v(x) = s(x)$. Then from (3.34) we have

$$s(x) \geqq \mathbf{E}_x g(x_{\sigma_\varepsilon}) \geqq s(x) - \varepsilon,$$

which proves the (ε,s)-optimality of the time σ_ε.

COROLLARY 2. *It follows from the proof of Lemma* 8 *that if the function* $g(x) \in L(A^+)$ *is such that* $g(x) \geqq K > -\infty$ *and it is bounded on the set* $(E\backslash\Gamma_0) \cup \partial\Gamma_0$, *then* $v(x) = \mathbf{E}_x v(x_{\sigma_\varepsilon})$.

By (3.34) it now follows that under these assumptions the time σ_ε is also (ε,s)-optima, $\varepsilon > 0$.

COROLLARY 3. *Suppose* $\hat{\mathfrak{M}} \subseteq \mathfrak{M}$ *is the class of stopping times* $\sigma_A = \inf\{t \geqq 0: x_t \in A\}$, *where* $A \in \hat{\mathscr{B}}$, *and*

$$\hat{s}(x) = \sup_{\sigma \in \hat{\mathfrak{M}}} \mathbf{E}_x g(x_\sigma).$$

If $g(x) \geqq K > -\infty$, *then* $s(x) = \tilde{s}(x) = \hat{s}(x)$.

In other words, it is sufficient to restrict consideration not to all Markov times but to only the times of first entry of almost Borel sets, in order to find the value in optimal stopping problems (under the assumption $g(k) \geqq K > -\infty$).

Indeed, if $|g(x)| \leqq K < \infty$, then the assertion follows from Corollary 1. In the general case we have to consider functions $g^b(x) = \min(g(x),b)$, where $b \geqq 0$, and then let $b \to \infty$.

COROLLARY 4. *Suppose the function* $g(x)$ *is continuous,* $g(x) \geqq K > -\infty$, *and the value* $s(x)$ *is lower semicontinuous (for this it is sufficient to require that* X *be a Feller process, according to Lemma* 2). *We let* $\hat{\mathfrak{M}}_D \subseteq \hat{\mathfrak{M}}$ *denote the class of times* $\sigma_D = \inf\{t \geqq 0: x_t \in D\}$, *where* D *are closed sets and*

$$\hat{s}_D(x) = \sup_{\sigma \in \hat{\mathfrak{M}}_D} \mathbf{E}_x g(x_\sigma).$$

Then $\bar{s}(x) = s(x) = \hat{s}(x) = \hat{s}_D(x)$.

For the proof it is sufficient to note that the sets $\Gamma_\varepsilon = \{x\colon s(x) \leqq g(x) + \varepsilon\}$ are closed.

5. In Lemma 9, given below, we weaken the assumption of boundedness of the function $g(x)$ made in Lemma 8, though at the cost of strengthening the smoothness requirements on the functions $g(x)$, $v_n(x) = \lim_{N\to\infty} Q_n^N g(x)$ and $v(x) = \lim_{n\to\infty} v_n(x)$.

LEMMA 9. *Suppose* $X = (x_t, \mathscr{F}_t, \mathbf{P}_x)$, $t \in \boldsymbol{T}$, $x \in E$, *is a standard Markov process, and* $g(x) \in L(A^-, A^+)$. *Assume that each of the functions* $g(x)$, $v_n(x)$, $n \in \boldsymbol{N}$, *is continuous and* $v_n(x) \to v(x)$ *as* $n \to \infty$, *uniformly in* x. *Then for every* $\varepsilon > 0$

$$v(x) = \mathbf{E}_x v(x_{\sigma_\varepsilon}). \tag{3.35}$$

If $\mathbf{P}_x(\sigma_0 < \infty) = 1$, $x \in E$. *then* $v(x) = \mathbf{E}_x v(x_{\sigma_0})$.

PROOF. Since the functions $v_n(x)$ are assumed to be continuous, we have $v_n(x) \to v(x)$ uniformly in x as $n \to \infty$ and $|v(x)| < \infty$, from which it follows that the function $v(x)$ is continuous ([33], Chapter VII, Theorem 6; [2], Chapter V, Theorem 22) and

$$\lim_{n\to\infty} v_n(x_n) = v(x), \tag{3.36}$$

if $x_n \to x$ as $n \to \infty$ ([33], Chapter VII, Theorem 2).

Let

$$\Gamma_\varepsilon^n = \{x\colon v_n(x) \leqq g(x) + \varepsilon\}, \qquad \sigma_\varepsilon^n = \inf\{k \cdot 2^{-n}\colon x_{k\cdot 2^{-n}} \in \Gamma_\varepsilon^n\},$$

where $k \in \boldsymbol{N}$.

Since $v_n(x) \uparrow v(x)$, it follows that $\Gamma_\varepsilon^n \downarrow \Gamma_\varepsilon$ as $n \to \infty$. Clearly $\sigma_\varepsilon \geqq \sigma_\varepsilon^n$ and $\sigma_\varepsilon^{n+1} \geqq \sigma_\varepsilon^n$ ($\mathbf{P}_x$-a.s., $x \in E$) and, by Lemma 7, $\mathbf{P}_x(\sigma_\varepsilon < \infty) = 1$, $x \in E$. Therefore $\sigma_\varepsilon^* = \lim_n \sigma_\varepsilon^n$ exists and $\sigma_\varepsilon \geqq \sigma_\varepsilon^*$ ($\mathbf{P}_x$-a.s., $x \in E$).

The functions $g(x)$, $v_n(x)$ and $v(x)$ are continuous. Hence it follows that the sets Γ_ε^n and Γ_ε are closed, $x_{\sigma_\varepsilon^n} \in \Gamma_\varepsilon^n$ and (by the right continuity of the process X) $x_{\sigma_\varepsilon} \in \Gamma_\varepsilon$, i.e.

$$v_n(x_{\sigma_\varepsilon^n}) \leqq g(x_{\sigma_\varepsilon^n}) + \varepsilon, \qquad v(x_{\sigma_\varepsilon}) \leqq g(x_{\sigma_\varepsilon}) + \varepsilon. \tag{3.37}$$

Since the process X is quasi-left-continuous, on passing to the limit ($n \to \infty$) in the first inequality in (3.37) we obtain

$$v(x_{\sigma_\varepsilon^*}) \leqq g(x_{\sigma_\varepsilon^*}) + \varepsilon, \tag{3.38}$$

from which it follows that $x_{\sigma^*_\varepsilon} \in \Gamma_\varepsilon$ and $\sigma^*_\varepsilon \geqq \sigma^*_\varepsilon$. Consequently $\sigma^*_\varepsilon = \sigma_\varepsilon$ ($\mathbf{P}_x$-a.s., $x \in E$).

Since $\mathbf{P}_x(\sigma_\varepsilon < \infty) = 1$, $x \in E$, and $\sigma^n_\varepsilon \leqq \sigma_\varepsilon$, by (2.25) we find

$$v_n(x) = \mathbf{E}_x v_n(x_{\sigma^n_\varepsilon}) \geqq \mathbf{E}_x v(x_{\sigma^n_\varepsilon}). \tag{3.39}$$

We shall show that

$$\lim_{x \to \infty} \mathbf{E}_x v(x_{\sigma^n_\varepsilon}) = \mathbf{E}_x v(x_{\sigma^n_\varepsilon}). \tag{3.40}$$

We have

$$\begin{aligned} \mathbf{E}_x v(x_{\sigma^n_\varepsilon}) &= \int_\Omega v(x_{\sigma^n_\varepsilon})\, d\mathbf{P}_x \\ &= \int_{\{\omega: \sigma_\varepsilon = \sigma^n_\varepsilon < \infty\}} v(x_{\sigma_\varepsilon})\, d\mathbf{P}_x + \int_{\{\omega: \sigma^n_\varepsilon < \sigma_\varepsilon < \infty\}} v(x_{\sigma^n_\varepsilon})\, d\mathbf{P}_x. \end{aligned}$$

But (cf. the proof of Lemma II.9)

$$\begin{aligned} \limsup_n & \int_{\{\omega:\, \sigma^n_\varepsilon < \sigma_\varepsilon < \infty\}} v(x_{\sigma^n_\varepsilon})\, d\mathbf{P}_x \\ & \leqq \limsup_n \int_{\{\omega:\, \sigma^n_\varepsilon < \sigma_\varepsilon < \infty\}} \sup_{n \geqq 0} g^+(x_n)\, d\mathbf{P}_x = 0, \\ \liminf_n & \int_{\{\omega:\, \sigma^n_\varepsilon < \sigma_\varepsilon < \infty\}} v(x_{\sigma^n_\varepsilon})\, d\mathbf{P}_x \\ & \geqq - \liminf_n \int_{\{\omega: \sigma^n_\varepsilon < \sigma_\varepsilon < \infty\}} \sup_{n \geqq 0} g^-(x_n)\, d\mathbf{P}_x = 0 \end{aligned}$$

and $\mathbf{E}_x \left| v(x_{\sigma_\varepsilon}) \right| < \infty$, $x \in E$. Consequently

$$\lim_{n \to \infty} \mathbf{E}_x v(x_{\sigma^n_\varepsilon}) = \lim_{n \to \infty} \int_{\{\omega:\, \sigma = \sigma^n_\varepsilon\}} v(x_{\sigma_\varepsilon})\, d\mathbf{P}_x = \mathbf{E}_x v(x_{\sigma_\varepsilon}).$$

From (3.39) and (3.40) we obtain the inequality

$$v(x) \leqq \mathbf{E}_x v(x_{\sigma_\varepsilon}), \qquad \varepsilon > 0,$$

which, as can be seen from the proof, also holds when $\varepsilon = 0$ if $\mathbf{P}_x(\sigma_0 < \infty) = 1$, $x \in E$. The reverse inequality follows from Lemma 5, and this proves the desired assertion (3.35).

COROLLARY 1. *If the conditions of the lemma are satisfied, then for all* $\varepsilon > 0$ *the time* σ_ε *is* (ε,s)*-optimal.*

Indeed, since the value $s(x)$ coincides with the s.e.m. $v(x)$ of the function $g(x)$, we see that for all $x \in E$

$$s(x) \geqq \mathbf{E}_x g(x_{\sigma_\varepsilon}) \geqq \mathbf{E}_x s(x_{\sigma_\varepsilon}) - \varepsilon = s(x) - \varepsilon.$$

COROLLARY 2. *Corollaries* 3 *and* 4 *of Lemma* 8 *hold under the assumptions of Lemma* 9.

6. It is now natural to investigate the structure of the value without the assumption $g \in L(A^-)$. Discussions analogous to those given at the beginning of § 3 of Chapter II for the case of discrete time show that if we do away with the assumption $g \in L(A^-)$, then the value $s(x)$ need no longer be the smallest excessive majorant function.

In order to state the basic results of this subsection and the following one (Theorems 2 and 3), we give some needed notation and definitions.

We let $\mathfrak{N}_g$ denote the class of stopping times $\tau \in \mathfrak{M}$ for which the mathematical expectation $\mathbf{E}_x g(x_\tau)$ is defined for all $x \in E$ $\big(g(x) \in L\big)$.

By analogy with the discrete time case (§ 3, Chapter II) we introduce the classes of times $\bar{\mathfrak{N}}_g, \mathfrak{R}_g$ and $\bar{\mathfrak{R}}_g$ and show[6] that in defining the values $\bar{s}(x)$ and $s(x)$ it suffices to take the supremum over the classes $\bar{\mathfrak{R}}_g$ and $\mathfrak{R}_g$ instead of $\bar{\mathfrak{N}}_g$ and $\mathfrak{N}_g$ respectively. Analogously,

$$\bar{s}(x) = \sup_{\tau \in \bar{\mathfrak{N}}_g} \mathbf{E}_x g(x_\tau) = \sup_{\tau \in \bar{\mathfrak{R}}_g} \mathbf{E}_x g(x_\tau).$$

DEFINITION 1. An almost Borel, $\mathscr{C}_0$-lower semicontinuous function $f(x) \in L$ is said to be $\bar{\mathfrak{N}}_g$*-regular* if $\tilde{\mathbf{E}}_x f(x_\tau)$ is defined for all $\tau \in \bar{\mathfrak{N}}_g$ and

$$f(x) \geqq \tilde{\mathbf{E}}_x f(x_\tau). \tag{3.41}$$

DEFINITION 2. An almost Borel, $\mathscr{C}_0$-lower semicontinuous function $f(x) \in L$ is said to be $\bar{\mathfrak{R}}_g$*-regular* if $\tilde{\mathbf{E}}_x f^-(x_\tau) < \infty$ for all $\tau \in \bar{\mathfrak{R}}_g$ and the inequality (3.41) is satisfied.

For brevity we call $\bar{\mathfrak{R}}_g$-regular functions simply *regular*.

Let $g_a(x) = \max\big(g(x), a\big)$, $a \leqq 0$,

$$v_n(x; a) = \lim_{N \to \infty} Q_n^N g_a(x), \tag{3.42}$$

$$v^*(x; a) = \lim_{n \to \infty} v_n(x; a) = \lim_{n \to \infty} \lim_{N \to \infty} Q_n^N g_a(x), \tag{3.43}$$

$$v^*(x) = \lim_{a \to -\infty} v^*(x; a). \tag{3.44}$$

THEOREM 2. *Suppose $X = (x_t, \mathscr{F}_t, \mathbf{P}_x)$, $t \in \boldsymbol{T}$, is a standard Markov process. Suppose $g(x) \in L(A^+)$, each of the functions $g(x)$, $v_n(x;a)$ and $v^*(x)$ is continuous in x (for all $a \leqq 0$ and $n \in \boldsymbol{N}$) and $v_n(x;a) \to v(x;a)$ as $n \to \infty$, uniformly in x for each $a \leqq 0$.*

Then the value $s(x)$ is an $\mathfrak{R}$-regular excessive majorant of the function $g(x)$ and $\bar{s}(x) = s(x)$ $\big(\bar{s}(x) = \bar{s}(x) = s(x)$ in the case $g(x) \geqq 0\big)$.

PROOF. By Theorem 1 the values

---

[6] We shall omit the subscript g on $\mathfrak{N}_g$, $\bar{\mathfrak{N}}_g$, $\mathfrak{R}_g$ and $\bar{\mathfrak{R}}_g$ whenever this does not give rise to ambiguity.

$$s_a(x) = \sup_{\tau \in \mathfrak{R}} \mathbf{E}_x g_a(x_\tau), \qquad \tilde{s}_a(x) = \sup_{\tau \in \tilde{\mathfrak{R}}} \tilde{\mathbf{E}}_x g_a(x_\tau)$$

coincide, and $s_a(x) = v^*(x;a)$. Since the $s_a(x)$ do not increase as $a \to -\infty$, it follows that $s_a(x) \downarrow v^*(x)$ and

$$v^*(x) \geqq \tilde{s}(x) \geqq s(x). \tag{3.45}$$

$\tilde{\mathbf{E}}_x v^*(x_\tau)$ is defined for every $\tau \in \tilde{\mathfrak{R}}$ and, according to Lemma 5,

$$s_a(x) \geqq \tilde{\mathbf{E}}_x s_a(x_\tau) \geqq \tilde{\mathbf{E}}_x v^*(x_\tau),$$

whence

$$v^*(x) \geqq \tilde{\mathbf{E}}_x v^*(x_\tau). \tag{3.46}$$

Thus the function

$$v^*(x) = \lim_{a \to -\infty} \lim_{n \to \infty} \lim_{N \to \infty} Q_n^N g_a(x) \tag{3.47}$$

is an $\tilde{\mathfrak{R}}$-regular excessive majorant of $g(x)$.

We shall establish the inequality $v^*(x) \leqq s(x)$, which together with (3.45) leads to the equalities $v^*(x) = \tilde{s}(x) = s(x)$.

As in the case of discrete time (see the proof of Lemma II.10), it can be shown that

$$\limsup_t g(x_t) = \limsup_t v^*(x_t) \tag{3.48}$$

and that for every $\varepsilon > 0$ the time

$$\sigma_\varepsilon^* = \inf\{t \geqq 0 \colon v^*(x_t) \leqq g(x_t) + \varepsilon\} \in \mathfrak{M}.$$

Since the functions $g(x)$ and $v^*(x)$ are continuous,

$$s(x) \geqq \mathbf{E}_x g(x_{\sigma_\varepsilon^*}) \geqq \mathbf{E}_x v^*(x_{\sigma_\varepsilon^*}) - \varepsilon. \tag{3.49}$$

Set $\Gamma_\varepsilon^*(a,b) = \{x \colon v^*(x,a) \leqq g(x,b) + \varepsilon\}$, $\Gamma_\varepsilon^*(-\infty,b) = \{x \colon v^*(x) \leqq g(x,b) + \varepsilon\}$ and $\sigma_\varepsilon^*(a,b) = \inf\{t \geqq 0 \colon x_t \in \Gamma_\varepsilon^*(a,b)\}$, $\sigma_\varepsilon^*(-\infty,b) = \inf\{t \geqq 0 \colon x_t \in \Gamma_\varepsilon^*(-\infty,b)\}$. Then as $a \to -\infty$, $\Gamma_\varepsilon^*(a,b) \uparrow \Gamma_\varepsilon^*(-\infty,b)$ and $\sigma_\varepsilon^*(a,b) \to \sigma_\varepsilon^*(-\infty,b)$ (by the closedness of the sets and the continuity of the process X from the right). Further, $\Gamma_\varepsilon^*(-\infty,b) \downarrow \Gamma_\varepsilon^* = \{x \colon v^*(x) \leqq g(x) + \varepsilon\}$ as $b \to -\infty$ and $\sigma_\varepsilon^*(-\infty,b) \geqq \sigma_\varepsilon^* = \inf\{t \geqq 0 \colon x_t \in \Gamma_\varepsilon^*\}$. Then by Lemmas 9 and 5,

$$\begin{aligned} v^*(x) \leqq v^*(x,b) &= \mathbf{E}_x v^*(x_{\sigma_\varepsilon^*(a,b)}, b) \\ &\leqq \mathbf{E}_x v^*(x_{\sigma_\varepsilon^*(-\infty,b)}, b) \leqq \mathbf{E}_x v^*(x_{\sigma_\varepsilon^*}, b). \end{aligned}$$

Hence by Fatou's Lemma

$$v^*(x) \leqq \limsup_{a\to-\infty} \mathbf{E}_x v^*(x_{\sigma_\varepsilon^*}; a) \leqq \mathbf{E}_x v^*(x_{\sigma_\varepsilon^*}), \tag{3.50}$$

which together with (3.49) leads to the inequality $v^*(x) \leqq s(x) + \varepsilon$. Since $\varepsilon > 0$ is arbitrary, it follows that $v^*(x) \leqq s(x)$, which together with (3.45) leads to the desired equalities $v^*(x) = \bar{s}(x) = s(x)$ $\big(= \bar{s}(x)$ if $g(x) \geqq 0\big)$.

Suppose $f(x)$ is an $\mathfrak{N}$-regular e.m. of the function $g(x)$. Then $f(x) \geqq \tilde{\mathbf{E}}_x f(x_\tau) \geqq \tilde{\mathbf{E}}_x g(x_\tau)$, whence it follows immediately that $f(x) \geqq \bar{s}(x)$. Thus $\bar{s}(x)$ is the smallest $\mathfrak{N}$-regular excessive majorant of the function $g(x)$. Theorem 2 is proved.

REMARK 1. Under the assumptions of Theorem 2 the value

$$s(x) = \lim_{a\to-\infty} \lim_{n\to\infty} \lim_{N\to\infty} Q_n^N g_a(x). \tag{3.51}$$

The proof follows from (3.47).

REMARK 2. Under the assumptions of Theorem 2 the value

$$\begin{aligned} s(x) &= \lim_{a\to-\infty} \lim_{n\to\infty} \lim_{b\to\infty} \lim_{N\to\infty} Q_n^N g_a^b(x) \\ &= \lim_{a\to-\infty} \lim_{n\to\infty} \lim_{N\to\infty} \lim_{b\to\infty} Q_n^N g_a^b(x), \end{aligned} \tag{3.52}$$

where $b \geqq 0$, $a \leqq 0$ and

$$g_a^b(x) = \begin{cases} b, & g(x) > b, \\ g(x), & a \leqq g(x) \leqq b, \\ a, & g(x) < a. \end{cases}$$

The proof follows from Remark 1 to Lemma 1, (3.51) and (3.44).

REMARK 3. Under the assumptions of Theorem 2 the value

$$s(x) = \lim_{a\to-\infty} \lim_{n\to\infty} \lim_{N\to\infty} G_{n,a}^N \varphi_{n,a}(x), \tag{3.53}$$

where

$$\varphi_{n,a}(x) = \mathbf{E}_x\Big\{\sup_{k\in N} g_a(x_{k\cdot 2^{-n}})\Big\},$$

$$G_{n,a}\varphi_{n,a}(x) = \max\{g_a(x), T_{2^{-n}}\varphi_{n,a}(x)\}$$

and $G_{n,a}^N$ is the Nth power of the operator $G_{n,a}$.

This follows from (3.18), (3.44) and (3.51).

7. We are now in a position to prove an assertion analogous to Theorem II.3. Let

$$v_n(x; a, b) = \lim_{N\to\infty} Q_n^N g_a^b(x),$$

$$v(x; a, b) = \lim_{n\to\infty} v_n(x; a, b),$$

$$v(x; b) = \lim_{a\to-\infty} v(x; a, b).$$

THEOREM 3. *Suppose* $X = (x_t, \mathscr{F}_t, \mathbf{P}_x)$, $t \in \boldsymbol{T}$, *is a standard Markov process and* $g(x)$ *is a continuous function from the class L. We assume that for all* $b \geqq 0$ *and* $a \leqq 0$ *the functions* $\upsilon_n(x;a,b)$, $\upsilon(x;a,b)$ *and* $\upsilon(x;b)$ *are continuous in x and* $\upsilon_n(x;a,b) \to \upsilon(x;a,b)$ *uniformly in x as* $n \to \infty$. *Then the value* $s(x)$ *is a regular excessive majorant of the function* $g(x)$ *and* $\bar{s}(x) = s(x)$ $\left(\tilde{s}(x) = \bar{s}(x) = s(x)\right.$ *in the case* $\left.g(x) \geqq 0\right)$.

PROOF. First of all we note that

$$s(x) = \sup_{\tau \in \mathfrak{R}} \mathbf{E}_x g(x_\tau), \qquad \bar{s}(x) = \sup_{\tau \in \bar{\mathfrak{R}}} \mathbf{E}_x g(x_\tau),$$
$$\tilde{s}(x) = \sup_{\tau \in \bar{\mathfrak{R}}} \tilde{\mathbf{E}}_x \, g(x_\tau).$$

Let

$$s^b(x) = \sup_{\tau \in \mathfrak{R}} \mathbf{E}_x g^b(x_\tau), \qquad \bar{s}^b(x) = \sup_{\tau \in \bar{\mathfrak{R}}} \mathbf{E}_x g^b(x_\tau),$$
$$\tilde{s}^b(x) = \sup_{\tau \in \bar{\mathfrak{R}}} \tilde{\mathbf{E}}_x g^b(x_\tau).$$

Clearly $s^b(x) \leqq s(x)$, $\lim_{b\to\infty} s^b(x)$ exists and $s^*(x) = \lim_{b\to\infty} s^b(x) \leqq s(x) \leqq \bar{s}(x)$.

On the other hand, analogously to (2.57) we can establish the inequality $\bar{s}(x) \leqq s^*(x)$, which together with the preceding inequalities shows that $s^*(x) = s(x) = \bar{s}(x)$.

Since, for $\tau \in \bar{\mathfrak{R}}$ we have $\mathbf{E}_x \bar{s}^-(x_\tau) \leqq \mathbf{E}_x g^-(x_\tau) < \infty$, $\bar{s}^b(x) \uparrow \bar{s}(x)$ and (by Theorem 2)

$$\bar{s}(x) \geqq \bar{s}^b(x) \geqq \mathbf{E}_x \bar{s}^b(x_\tau), \tag{3.54}$$

on passing to the limit $(b \to \infty)$ in (3.54) we obtain

$$\bar{s}(x) \geqq \mathbf{E}_x \bar{s}(x_\tau). \tag{3.55}$$

The function $s(x) = \lim_{b\to\infty} s^b(x)$ is lower semicontinuous $\left(\text{since } s^b(x) \text{ is continuous in } x \text{ and } s^b(x) \uparrow s(x)\right)$ and consequently it is almost Borel and $\mathscr{C}_0$-lower semicontinuous. Together with (3.55) this shows that $s(x)$ is a regular (more precisely, $\bar{\mathfrak{R}}_g$-regular) excessive majorant of the function $g(x)$. It is also obvious that the function $s(x)$ is the smallest e.m. of $g(x)$. Theorem 3 is proved.

REMARK 1. Suppose the assumptions of Theorem 3 are satisfied. Then

$$s(x) = \lim_{b\to\infty} \lim_{a\to-\infty} \lim_{n\to\infty} \lim_{N\to\infty} Q_n^N g_a^b(x). \tag{3.56}$$

REMARK 2. Remark 2 to Theorem II.3 and Corollaries 3 and 4 of Lemma 8 remain valid with the obvious changes in formulation and notation.

8. As in § 8 of Chapter II, we consider some other formulations of optimal stopping problems.

Let

$$s(x) = \sup \mathbf{E}_x\left\{e^{-\lambda\tau}g(x_\tau) - \int_0^\tau e^{-\lambda s}c(x_s)ds\right\}, \tag{3.57}$$

where the functions $g \in L$, $c \in L$, $\lambda \geqq 0$, and the upper bound is taken over the class $\mathfrak{N} = \{\tau\}$ of stopping times $\tau = \tau(\omega)$ for which the mathematical expectation in (3.57) is defined.

In many cases the problem of finding the value defined in (3.57) can be reduced to solving a new problem in which the fee $c(x) \equiv 0$. We investigate several of these.

We assume that

$$\mathbf{E}_x \int_0^\infty e^{-\lambda s} \mid c(x_s) \mid ds < \infty, \qquad x \in E,$$

and let

$$f(x) = \mathbf{E}_x \int_0^\infty e^{-\lambda s}c(x_s)ds.$$

Then it is known (see Theorem 5.1 in [25]) that for every stopping time $\tau = \tau(\omega)$

$$\mathbf{E}_x[e^{-\lambda\tau}f(x_\tau)] - f(x) = -\mathbf{E}_x \int_0^\tau e^{-\lambda s}c(x_s)ds.$$

Consequently

$$\begin{aligned} s(x) &= \sup_{\tau\in\mathfrak{N}} \mathbf{E}_x\left\{e^{-\lambda\tau}g(x_\tau) - \int_0^\tau e^{-\lambda s}c(x_s)ds\right\} \\ &= \sup_{\tau\in\mathfrak{N}} \mathbf{E}_x\{e^{-\lambda\tau}G(x_\tau)\} - f(x), \end{aligned}$$

where $G(x) = g(x) + f(x)$ (cf. Remark 1 to Theorem II.16).

Thus if $\mathbf{E}_x \int_0^\infty e^{-\lambda s}\left| c(x_s) \right| ds < \infty$, $x \in E$, then to find the value $s(x)$ it is sufficient to be able to find

$$S(x) = \sup_{\tau\in\mathfrak{N}} \mathbf{E}_x e^{-\lambda\tau}G(x_\tau),$$

since $s(x) = S(x) - f(x)$.

By introducing the new variable $x' = (x,t)$, we can reduce solution of the given problem to the one already considered above, taking the function $g'(x') = e^{-\lambda t}G(x)$ as the reward. We can also proceed in another fashion.

We say that the $\mathscr{C}_0$-continuous function $F(x)$ is a *λ-excessive majorant of the function* $G(x)$ if $F(x) \geqq G(x)$ and

$$e^{-\lambda t}T_tF(x) \leqq F(x).$$

From Theorem 1 it is easy to derive (or prove by the same method) that under the assumption

$$A^-: \mathbf{E}_x\Big\{\sup_t [e^{-\lambda t} g^-(x_t)]\Big\} < \infty, \qquad x \in E,$$

the value $S(x)$ is the smallest λ-excessive majorant of the function $G(x)$. Here

$$S(x) = \lim_{n\to\infty} \lim_{N\to\infty} Q_n^N G(x),$$

where

$$Q_n G(x) = \max\{G(x), e^{-\lambda\Delta} T_\Delta G(x)\}, \qquad \Delta = 2^{-n}.$$

If the conditions of Theorem 3 are satisfied (with the obvious changes in notation), then the value $S(x)$ is the smallest regular λ-excessive majorant of the function $G(x)$.

In those cases in which the condition A^- is violated, we can give the following useful approach to reducing problems with a fee to the case $c(x) \equiv 0$.

Suppose $|c(x)| \leqq C < \infty$ and $f(x)$ is a (bounded) solution of the equation $\tilde{\mathscr{A}} f(x) = -c(x)$, where $\tilde{\mathscr{A}}$ is the weak infinitesimal generator of the process X. We let $\mathfrak{M}^*$ denote the class of those stopping times $\tau = \tau(\omega)$ for which $\mathbf{E}_x \tau < \infty$, $x \in E$. From the Corollary to Theorem 5.1 in [25] we have

$$\mathbf{E}_x f(x_\tau) - f(x) = -\mathbf{E}_x \int_0^\tau c(x_s) ds. \tag{3.58}$$

Therefore, if

$$s^*(x) = \sup_{\tau\in\mathfrak{M}^*} \mathbf{E}_x\left[g(x_\tau) - \int_0^\tau c(x_s)ds\right],$$

then

$$s^*(x) = \sup_{\tau\in\mathfrak{M}^*} \mathbf{E}_x[G(x_\tau)] - f(x),$$

where $G(x) = g(x) + f(x)$.

The validity of (3.58) can also be established without the assumption of boundedness of the function $g(x)$ (see, for example, the problems considered below in Chapter IV), which allows us to make use of the theory developed above for solving problems with a fee.

If the function $G(x)$ satisfies the condition

$$A^-: \mathbf{E}_x\left[\sup_t G^-(x_t)\right] < \infty, \qquad x \in E$$

(see also Remark 4 to Lemma II.4), then by Lemma III.1

$$s^*(x) + f(x) = \lim_{n\to\infty} \lim_{N\to\infty} Q_n^N G(x),$$

where

$$Q_n G(x) = \max\{G(x),\ P_\Delta G(x)\}, \qquad \Delta - 2^{-n}.$$

We set

$$Q_{n,c}g(x) \max\left\{g(x),\ -\mathbf{E}_x \int_0^\Delta c(x_s)ds + T_\Delta g(x)\right\}, \qquad \Delta = 2^{-n}.$$

Then it can easily be seen that by (3.58)

$$\begin{aligned} Q_n G(x) &= \max\{g(x) + f(x),\ T_\Delta[g(x) + f(x)]\} \\ &= \max\left\{g(x) + f(x), f(x) - \mathbf{E}_x \int_0^\Delta c(x_s)ds + T_\Delta g(x)\right\} \\ &= f(x) + \max\left\{g(x),\ -\mathbf{E}_x \int_0^\Delta c(x_s)ds + T_\Delta g(x)\right\} \\ &= f(x) + Q_{n,c}g(x). \end{aligned}$$

Therefore

$$s(x) = \lim_{n\to\infty} \lim_{N\to\infty} Q_{n,c}^N g(x).$$

§ 4. ε-optimal and optimal Markov times

1. THEOREM 4. *Suppose X is a standard process, the function $g(x) \in L(A^+)$ is bounded from below, and is bounded on the closed set $E\backslash\Gamma_0 = \{x\colon v(x) > g(x)\}$, where $v(x)$ is the s.e.m. of $g(x)$.*

1. *For every $\varepsilon > 0$ the time*

$$\sigma_\varepsilon = \inf\{t \geqq 0\colon v(x_t) \leqq g(x_t) + \varepsilon\}$$

is an (ε,s)-optimal stopping time.

2. *If the function $g(x)$ is continuous and $v(x)$ is lower semicontinuous, then the time*

$$\sigma_0 = \inf\{t \geqq 0\colon v(x_t) = g(x_t)\}$$

is $(0,\bar{s})$-optimal.

3. *If $g(x)$ is continuous, $v(x)$ is lower semicontinuous and the time $\sigma_0 \in \mathfrak{M}$, then σ_0 is an optimal stopping time.*

PROOF. Assertion 1 follows immediately from Corollary 2 of Lemma 8. For the proof of the remaining assertions we need

LEMMA 10. *Suppose X is a standard process, the function $g(x)$ is continuous, and* *the function $f(x)$ is lower semicontinuous. Let $\sigma_\varepsilon = \inf\{t \geqq 0\colon x_t \in \Gamma_\varepsilon\}$, where $\Gamma_\varepsilon = \{x\colon f(x) \leqq g(x) + \varepsilon\}$, $\varepsilon \geqq 0$. Then the sets Γ_ε are closed, $\Gamma_\varepsilon \downarrow \Gamma_0$ as $\varepsilon \downarrow 0$, and $\sigma_\varepsilon \uparrow \sigma_0$ ($\mathbf{P}_x$-a.s., $x \in E$).*

PROOF. If $x_n \to x$ as $n \to \infty$, where $x_n \in \Gamma_\varepsilon$, then

$$f(x) \leqq \liminf_{n\to\infty} f(x_n) \leqq \liminf_{n\to\infty} g(x_n) + \varepsilon = \lim_{n\to\infty} g(x_n) + \varepsilon = g(x) + \varepsilon,$$

whence it follows that the set Γ_ε is closed. It is also obvious that $\Gamma_\varepsilon \downarrow \Gamma_0$ as $\varepsilon \downarrow 0$, and the limit $\lim_{\varepsilon\downarrow 0}\sigma_\varepsilon = \sigma$ exists for all $x \in E$ with $\mathbf{P}_x$-probability one. We shall show that $\sigma = \sigma_0$ ($\mathbf{P}_x$-a.s., $x \in E$). Since $\sigma_0 \geqq \sigma_\varepsilon$, we have $\sigma_0 \geqq \sigma$. If $\sigma(\omega) = \infty$, then $\sigma_0(\omega) = \infty$ and consequently $\sigma(\omega) = \sigma_0(\omega)$ on the set $A = \{\omega\colon \sigma(\omega) = \infty\}$. Now suppose $\omega \in \Omega \setminus A$. In view of the quasi-left-continuity of the process X, it follows that $x_{\sigma_\varepsilon} \to x_\sigma$ ($\mathbf{P}_x$-a.s., $x \in E$) on the set $\Omega \setminus A$. Since the process X is continuous from the right and the set Γ_ε is closed, we have $x_{\sigma_\varepsilon} \in \Gamma_\varepsilon$. Consequently $f(x_{\sigma_\varepsilon}) \leqq g(x_{\sigma_\varepsilon}) + \varepsilon$, and as $\varepsilon \to 0$ on the set $\Omega \setminus A$ we have

$$f(x_\sigma) \leqq \liminf_{\varepsilon\downarrow 0} f(x_{\sigma_\varepsilon}) \leqq \liminf_{\varepsilon\downarrow 0}[g(x_{\sigma_\varepsilon}) + \varepsilon] = g(x_\sigma) \qquad (\mathbf{P}_x\text{-a.s., } x \in E). \tag{3.59}$$

It follows from (3.59) that $x_\sigma \in \Gamma_0$, and thus $\sigma_0(\omega) \leqq \sigma(\omega)$, $\omega \in \Omega \setminus A$, and therefore $\sigma_0(\omega) = \sigma(\omega)$ ($\mathbf{P}_x$-a.s., $x \in E$).

We proceed with the proof of assertions 2 and 3 of Theorem 4. By Lemma 10, $\sigma_\varepsilon \uparrow \sigma_0$ as $\varepsilon \downarrow 0$. According to (3.34), for $\varepsilon > 0$

$$v(x) - \varepsilon \leqq \int_{(\sigma_0<\infty)} g(x_{\sigma_\varepsilon})d\mathbf{P}_x + \int_{(\sigma_0=\infty,\ \sigma_\varepsilon<\infty)} g(x_{\sigma_\varepsilon})d\mathbf{P}_x. \tag{3.60}$$

Passing to the limit as $\varepsilon \to 0$ in (3.60), we obtain by Fatou's Lemma

$$v(x) \leqq \int_{(\sigma_0<\infty)} g(x_{\sigma_0})d\mathbf{P}_x + \int_{(\sigma_0=\infty)} \limsup_{t\to\infty} g(x_t)d\mathbf{P}_x = \tilde{\mathbf{E}}_x g(x_{\sigma_0}), \tag{3.61}$$

which proves assertion 2.

If $\mathbf{P}_x(\sigma_0 = \infty) = 0$, then in view of the conditions $g(x) \geqq C > -\infty$ and $g(x) \in L(A^+)$ we have

$$\int_{(\sigma_0=\infty)} \limsup_{t\to\infty} g(x_t)d\mathbf{P}_x = 0,$$

from which it follows that σ_0 is a $(0,s)$-optimal stopping time.

REMARK 1. If X is a standard Markov process with a finite number of states and $-\infty < g(x) < \infty$, then an optimal stopping time always exists.

REMARK 2. If the continuous function $|g(x)| \leqq C < \infty$ and the standard process X is a Feller process, then the time

$$\sigma_0 = \inf\{t \geqq 0 : v(x_t) = g(x_t)\}$$

is $(0,\bar{s})$-optimal; σ_0 is a $(0,s)$-optimal time if $\mathbf{P}_x(\sigma_0 < \infty) = 1$, $x \in E$.

REMARK 3. Suppose that all the conditions of Theorem 4 are satisfied with the exception of the requirement $g(x) \in L(A^+)$. Suppose however that there exists a point x_0 such that $\mathbf{E}_{x_0}[\sup_t g^+(x_t)] < \infty$. Then the assertion of the theorem remains valid if by (ε,s)- and $(\varepsilon,\bar{s})$-optimality we mean the corresponding optimality at the point x_0.

2. The next theorem below is the analog of Theorem II.4 for the case of continuous time.

THEOREM 5. *Suppose the conditions of Theorem 2 are satisfied and $v(x)$ is the smallest $\mathfrak{N}_g$-regular e.m. of the function $g(x)$.*

1. *For each $\varepsilon > 0$ the time*

$$\sigma_\varepsilon = \inf\{t \geqq 0 : v(x_t) \leqq g(x_t) + \varepsilon\}$$

is (ε,s)-optimal.

2. *The time*

$$\sigma_0 = \inf\{t \geqq 0 : v(x_t) = g(x_t)\}$$

is $(0,\bar{s})$-optimal.

3. *If $\mathbf{P}_x(\sigma_0 < \infty) = 1$, $x \in E$, then the time σ_0 is $(0,s)$-optimal.*

PROOF. The first assertion follows from (3.19) and (3.50). For the proof of $(0,\bar{s})$-optimality of the time σ_0 we make use of the inequality (3.50):

$$v(x) \leqq \mathbf{E}_x v(x_{\sigma_\varepsilon}) = \int_{(\sigma_0<\infty)} v(x_{\sigma_\varepsilon}) d\mathbf{P}_x + \int_{(\sigma_0=\infty,\ \sigma_\varepsilon<\infty)} v(x_{\sigma_\varepsilon}) d\mathbf{P}_x. \tag{3.62}$$

As in the proof of the inequality (2.61), we derive from this that

$$v(x) \leqq \int_{(\sigma_0<\infty)} v(x_{\sigma_\varepsilon}) d\mathbf{P}_x + \int_{(\sigma_0=\infty,\ \sigma_\varepsilon<\infty)} \sup_{t \geqq \sigma_\varepsilon} g(x_t) d\mathbf{P}_x. \tag{3.63}$$

Since the process X is quasi-left-continuous, by Fatou's Lemma (3.63) implies

$$\begin{aligned} v(x) &\leqq \int_{(\sigma_0<\infty)} \limsup_{\varepsilon \downarrow 0} v(x_{\sigma_\varepsilon}) d\mathbf{P}_x \\ &+ \int_{(\sigma_0=\infty)} \limsup_t g(x_t) d\mathbf{P}_x = \int_{(\sigma_0<\infty)} v(x_{\sigma_0}) d\mathbf{P}_x \\ &+ \int_{(\sigma_0=\infty)} \limsup_t g(x_t) d\mathbf{P}_x. \end{aligned}$$

But $v(x_{\sigma_0}) = g(x_{\sigma_0})$ on the set $(\sigma_0 < \infty)$. Consequently

$$v(x) \leqq \int_{(\sigma_0<\infty)} g(x_{\sigma_0})d\mathbf{P}_x + \int_{(\sigma_0=\infty)} \limsup_t g(x_t)d\mathbf{P}_x = \tilde{\mathbf{E}}_x g(x_{\sigma_0}),$$

which proves the second assertion of the theorem.

From the preceding inequality it follows that

$$v(x) \leqq \int_{(\sigma_0<\infty)} g(x_{\sigma_0})d\mathbf{P}_x + \int_{(\sigma_0=\infty)} \limsup_t g^+(x_t)d\mathbf{P}_x.$$

Therefore if $\mathbf{P}_x\{\sigma_0 = \infty\} = 0$, then

$$v(x) \leqq \int_\Omega g(x_{\sigma_0})d\mathbf{P}_x = \mathbf{E}_x g(x_{\sigma_0}).$$

which proves the $(0,s)$-optimality of the time σ_0. The theorem is proved.

COROLLARY. *If* $\lim_{t\to\infty} g(x_t) = -\infty$ ($\mathbf{P}_x$-*a.s.*, $x \in E$), *then the time* σ_0 *is* $(0,s)$-*optimal* (cf. the Corollary to Theorem II.4).

REMARK 1. Suppose $\mathfrak{M}^* = \{\tau\colon \mathbf{E}_x\tau < \infty, x \in E\}$ and

$$s(x) = \sup_{\tau\in\mathfrak{M}^*} \mathbf{E}_x\left[g(x_\tau) - \int_0^\tau c(x_s)ds\right],$$

where the functions $g(x)$ and $c(x)$ are continuous, $|g(x)| \leqq K < \infty$, $c(x) \geqq 0$ and

$$\int_0^\infty c(x_s)ds = \infty \qquad (\mathbf{P}_x\text{-a.s., } x \in E).$$

We also assume that there exists a function $f(x)$ (see § 3.8) such that for every $\tau \in \mathfrak{M}^*$

$$\mathbf{E}_x f(x_\tau) - f(x) = -\mathbf{E}_x \int_0^\tau c(x_s)ds.$$

Let $G(x) = g(x) + f(x)$; then obviously

$$s(x) = \sup_{\tau\in\mathfrak{M}^*} \mathbf{E}_x G(x_\tau) - f(x).$$

Suppose the function $G(x)$ and the functions $v_n(x;a,b)$, $v(x;a,b)$ and $v(x;b)$ (see (3.42)—(3.44)) constructed from it satisfy the conditions of Theorem 3. Then the time

$$\sigma_0 = \inf\{t \geqq 0\colon s(x_t) = g(x_t)\}$$

is an optimal stopping time.

REMARK 2. Remarks 1—3 to Theorem 4 remain valid, with obvious changes in notation.

3. It was shown in the case of discrete time (Theorem II.7) that essentially the class of optimal stopping times (if they exist) is exhausted by the times

$$\sigma_0 = \inf\{t \geqq 0: v(x_t) = g(x_t)\}.$$

The following theorem contains the corresponding result for continuous time.

THEOREM 6. *Suppose the almost Borel, $\mathscr{C}_0$-continuous function $g(x) \in L(A^-)$ and $s(x)$ is the value. Assume that at a point x_0 where the value $s(x_0) < \infty$ there exists a $(0,\tilde{s})$-optimal time τ^* (i.e., $\mathbf{E}_{x_0} g(x_{\tau^*}) = s(x_0)$). Then the time*

$$\sigma_0 = \inf\{t \geqq 0: s(x_t) = g(x_t)\}$$

is $(0,\tilde{s})$-optimal at the point x_0, and $\mathbf{P}_{x_0}\{\sigma_0 \leqq \tau^*\} = 1$. If, moreover, τ^* is a $(0,s)$-optimal stopping time, then the time σ_0 is also a $(0,s)$-optimal stopping time.

PROOF. Since $g(x) \in L(A^-)$, it follows that $s(x) \in L(A^-)$ and by Lemma 5

$$s(x) \geqq \tilde{\mathbf{E}}_x s(x_{\tau^*}) \geqq \tilde{\mathbf{E}}_x g(x_{\tau^*}). \tag{3.64}$$

But $\tilde{\mathbf{E}}_{x_0} g(x_{\tau^*}) = s(x_0)$ by hypothesis, and therefore from (3.64) we obtain

$$s(x_0) = \tilde{\mathbf{E}}_{x_0} g(x_{\tau^*}) = \tilde{\mathbf{E}}_{x_0} s(x_{\tau^*}).$$

We shall show that $s(x_{\tau^*}) = g(x_{\tau^*})$ ($\mathbf{P}_{x_0}$-a.s.) on the set $\{\tau^* < \infty\}$. It is clear that $s(x_{\tau^*}) \geqq g(x_{\tau^*})$ on the set $\{\tau^* < \infty\}$. Let

$$\mathbf{P}_{x_0}\{(\tau^* < \infty) \cap (s(x_{\tau^*}) > g(x_{\tau^*}))\} > 0.$$

Then

$$s(x_0) = \tilde{\mathbf{E}}_{x_0} s(x_{\tau^*}) > \tilde{\mathbf{E}}_{x_0} g(x_{\tau^*}) = s(x_0),$$

which is impossible if $s(x_0) < \infty$. Consequently $s(x_{\tau^*}) = g(x_{\tau^*})$ ($\mathbf{P}_x$-a.s.) on the set $\{\tau^* < \infty\}$.

In precisely the same way $\limsup_t s(x_t) = \limsup_t g(x_t)$ on the set $\{\tau^* = \infty\}$, and obviously $\sigma_0(\omega) \leqq \tau^*(\omega)$.

Thus $\sigma_0(\omega) \leqq \tau^*(\omega)$ ($\mathbf{P}_{x_0}$-a.s.), whence, taking into account that $x_{\sigma_0} \in \Gamma_0 = \{x: s(x) = g(x)\}$ and applying Lemma 5, we obtain

$$\begin{aligned}\tilde{\mathbf{E}}_{x_0} g(x_{\sigma_0}) &= \int_{(\sigma_0<\infty)} g(x_{\sigma_0}) d\mathbf{P}_{x_0} + \int_{(\sigma_0=\infty)} \limsup_t g(x_t) d\mathbf{P}_{x_0} \\ &= \int_{(\sigma_0<\infty)} s(x_{\sigma_0}) d\mathbf{P}_{x_0} + \int_{(\sigma_0=\infty)} \limsup_t s(x_t) d\mathbf{P}_{x_0} \\ &= \tilde{\mathbf{E}}_{x_0} s(x_{\sigma_0}) \geqq \tilde{\mathbf{E}}_{x_0} s(x_{\tau^*}) = s(x_0),\end{aligned}$$

i.e. the time σ_0 is $(0,\tilde{s})$-optimal.

If $\mathbf{P}_{x_0}\{\tau^* < \infty\} = 1$, then it can be shown analogously that $\mathbf{P}_{x_0}$-a.s. $\sigma_0 \leqq \tau^*$ and $\mathbf{E}_{x_0}g(x_{\sigma_0}) \geqq s(x_0)$. The theorem is proved.

§ 5. Integral and "differential" equations for the value

1. In the case of discrete time, $\boldsymbol{N} = \{0,1,\cdots\}$, the value $s(x)$ satisfied the recursion equation

$$s(x) = \max\{g(x),\ Ts(x)\}, \tag{3.65}$$

which yields a powerful method for finding it. It is natural to want to obtain an analog of this equation for the case of continuous time.

If follows from Theorems 1 – 3 that for all $t \geqq 0$

$$s(x) \geqq \max\{g(x),\ T_t s(x)\} \tag{3.66}$$

and, generally speaking, we need not have equality in (3.66) for any $t > 0$.

A different approach, which is simpler to explain first for the case of discrete times, turns out to be more productive.

LEMMA 11. *Suppose $X = (x_n, \mathscr{F}_n, \mathbf{P}_x)$, $n \in \boldsymbol{N}$, is a Markov chain with values in the phase space $(E,\mathscr{B})$. Let the function $g(x) \in L$ and suppose $f(x)$ is an excessive majorant of it satisfying the equation*

$$f(x) = \max\{g(x),\ Tf(x)\}. \tag{3.67}$$

Let $\Gamma_0 = \{x: f(x) = g(x)\}$ and, if V is a Borel set,

$$\sigma(V) = \inf\{n \geqq 0:\ x_n \in E \setminus V\}, \qquad \sigma_0 = \sigma(E \setminus \Gamma_0).$$

Then for all $N \in \boldsymbol{N}$ and $x \in E$

$$f(x) = \int_{(\sigma_0 \wedge \sigma(V) \leqq N)} f(x_{\sigma_0 \wedge \sigma(V)})\, d\mathbf{P}_x + \int_{(\sigma_0 \wedge \sigma(V) > N)} f(x_N)\, d\mathbf{P}_x, \tag{3.68}$$

where $\sigma_0 \wedge \sigma(V) = \min(\sigma_0,\ \sigma(V))$.

The proof is carried out exactly as the proof of Lemma II.5 We need only note that $\{\sigma_0 \wedge \sigma(V) > k\}$ on the set $f(x_k) = Tf(x_k)$.

COROLLARY 1. *Suppose the function $f \in L(A^-,A^+)$. Then*

$$f(x) = \tilde{\mathbf{E}}_x f(x_{\sigma_0 \wedge \sigma(V)}), \qquad x \in E. \tag{3.69}$$

If in particular $V \cap \Gamma_0 = \phi$, then

$$f(x) = \tilde{\mathbf{E}}_x f(x_{\sigma(V)}), \qquad x \in E. \tag{3.70}$$

COROLLARY 2. *Suppose $f \in L(A^-,A^+)$ and $\mathbf{P}_{x_0}(\sigma_0 \wedge \sigma(V) < \infty) = 1$; then*

$$f(x_0) = \mathbf{E}_{x_0} f(x_{\sigma_0 \wedge \sigma(V)}), \tag{3.71}$$

and if $V \cap \Gamma_0 = \phi$, *then*

$$f(x_0) = \mathbf{E}_{x_0} f(x_{\sigma(V)}). \tag{3.72}$$

Lemma 11 allows us to obtain a different form of the recursion equations for $f(x)$ which can be generalized easily to the case of continuous time.

LEMMA 12. *Assume that the conditions of Lemma* 11 *are satisfied and, moreover, the function* $g \in L(A^-, A^+)$. *For each* $x \in E$ *let* $V = V(x)$ *denote a Borel set containing the point* x. *Then*

$$f(x) = \max\{g(x), \mathbf{E}_x f(x_{\sigma_0 \wedge \sigma(V)})\}. \tag{3.73}$$

PROOF. If $x \in \Gamma_0$, then $f(x) = g(x)$ and (3.73) is obviously satisfied. But if $x \notin \Gamma_0$, then $f(x) > g(x)$ and by (3.69) the relationship (3.73) is again satisfied.

COROLLARY 1. *If the neighborhood* $V = V(x_0)$ *of the point* x_0 *is such that* $\inf_{x \in V(x_0)} \mathbf{P}_x\{\sigma(V) < \infty\} = 1$, *then for all* $x \in V(x_0)$, (3.73) *can also be written as follows*:

$$f(x) = \max\{g(x), \mathbf{E}_x f(x_{\sigma_0 \wedge \sigma(V)})\}. \tag{3.74}$$

Suppose further that for each Markov time $\tau \in \mathfrak{M}$ ($\tau \in \overline{\mathfrak{M}}$) we set $T_\tau f(x) = \mathbf{E}_x f(x_\tau)$ $(\tilde{T}_\tau f(x) = \tilde{\mathbf{E}}_x f(x_\tau))$. Then, for example, equation (3.74) can be rewritten as follows:

$$f(x) = \max\{g(x), \tilde{T}_{\sigma_0 \wedge \sigma(V)} f(x)\}. \tag{3.75}$$

2. Indeed, equations (3.73) and (3.74) (but not equation (3.67)) can be generalized to the case of continuous time.

THEOREM 7. *Suppose* X *is a standard Markov process. The function* $g \in L(A^+)$ *is nonnegative, continuous, and bounded on the closure of the set* $E \backslash \Gamma_0 = \{x: s(x) > g(x)\}$, *where* $s(x)$ *is the lower semicontinuous s.e.m. of the function* g(x).

Let $V = V(x)$ *denote a neighborhood* (with *compact closure*) *of the point* $x \in E$. *Then*

$$s(x) = \max\{g(x), \tilde{T}_{\sigma_0 \wedge \sigma(V)} s(x)\}, \tag{3.76}$$

where

$$\sigma(V) = \inf\{t \geqq 0: x_t \notin V\}, \qquad \sigma_0 = \sigma(E \backslash \Gamma_0).$$

PROOF. First we note that $\sigma(V)$ is a Markov time, since, by hypothesis, V is a neighborhood with compact closure ([**25**], Chapter IV, § 4.5). The set $E \backslash \Gamma_0$ is Borel, and consequently both σ_0 and $\sigma_0 \wedge \sigma(V)$ are also Markov times.

It was shown in Theorem 4 (see (3.61)) that $s(x) \leqq \tilde{\mathbf{E}}_x g(x_{\sigma_0})$. Therefore $s(x) \leqq \tilde{\mathbf{E}}_x s(x_{\sigma_0})$, and $\tilde{\mathbf{E}}_x s(x_{\sigma_0}) \leqq \tilde{\mathbf{E}}_x s(x_{\sigma_0 \wedge \sigma(V)}) \leqq s(x)$ by Lemma 5.

Consequently $s(x) = \tilde{\mathbf{E}}_x s(x_{\sigma_0 \wedge \sigma(V)})$ for all $x \in E$. Hence (as in Lemma 12) we immediately obtain equation (3.73). The theorem is proved.

Properly speaking, below it will not be the "integral" equation (3.73) itself which is important, but the relationship

$$s(x) = \tilde{\mathbf{E}}_x s(x_{\sigma_0 \wedge \sigma(V)}) \tag{3.77}$$

(of which (3.73) is a consequence), which allows us to obtain "differential" equations for the value (see (3.80)). We note that if the neighborhood $V = V(x_0)$ (having compact closure) is such that $V \cap \Gamma_0 = \phi$, then

$$s(x) = \tilde{\mathbf{E}}_x s(x_{\sigma(V)}) \tag{3.78}$$

and in particular, if $\inf_{x \in V(x_0)} \mathbf{P}_x\{\sigma(V) < \infty\} = 1$, then

$$s(x) = \mathbf{E}_x s(x_{\sigma(V)}) \tag{3.79}$$

for all points $x \in V(x_0)$.

REMARK 1. Suppose the function $G(x)$ introduced in Remark 1 to Theorem 5 satisfies the conditions of Theorem 7. Then the value

$$s(x) = \sup_{\tau \in \mathfrak{M}^*} \mathbf{E}_x \left[g(x_\tau) - \int_0^\tau c(x_s) ds \right],$$

where $\mathfrak{M}^* = \{\tau : \mathbf{E}_x \tau < \infty,\ x \in E\}$, $|g(x)| \leqq K < \infty$ and $c(x) \geqq 0$, satisfies the equation

$$s(x) = \max\left\{ g(x),\ -\mathbf{E}_x \int_0^{\sigma_0 \wedge \sigma(V)} c(x_s) ds + \tilde{T}_{\sigma_0 \wedge \sigma(V)} s(x) \right\}.$$

If, in particular, $V \cap \Gamma_0 = \phi$ and $\inf_{x \in V(x_0)} \mathbf{P}_x\{\sigma(V) < \infty\} = 1$, then

$$s(x) = \max\left\{ g(x),\ -\mathbf{E}_x \int_0^{\sigma(V)} c(x_s) ds + T_{\sigma(V)} s(x) \right\}.$$

The following important result can be derived from Theorem 7.

THEOREM 8. *Suppose the conditions of Theorem 7 are satisfied. Then the value belongs to the domain of definition $\mathscr{D}_{\mathfrak{A}}(E \setminus \Gamma_0)$ of the characteristic operator $\mathfrak{A}$ and satisfies the relationships*

$$\mathfrak{A} s(x) = 0, \qquad x \in C_0 = E \setminus \Gamma_0, \tag{3.80}$$

$$s(x) = g(x), \qquad x \in \Gamma_0. \tag{3.81}$$

The proof follows from the definition of the characteristic operator $\mathfrak{A}$ (see § 3 of Chapter I), Theorem 7, (3.79), and the fact that for each point $x \in E \setminus \Gamma_0$ there exists a neighborhood $V \subseteq E \setminus \Gamma_0$ having compact closure with $\mathbf{P}_x(\sigma(V) < \infty) = 1$. Indeed, suppose $x \in E \setminus \Gamma_0$. Then there is an $\varepsilon > 0$ such that $x \in E \setminus \Gamma_\varepsilon$, where

$\Gamma_\varepsilon = \{x: g(x) + \varepsilon \geqq s(x)\}$. The set Γ_ε is closed, and consequently $E \setminus \Gamma_\varepsilon$ is open. By the local compactness of the space E there is an open neighborhood U of the point x having compact closure $\bar{U}$. Consequently $V = (E \setminus \Gamma_\varepsilon) \cap U$ is an open neighborhood of the point x having compact closure. It is obvious that $\sigma(V) \leqq \sigma(E \setminus \Gamma_\varepsilon) = \sigma_\varepsilon$. Hence $\mathbf{P}_x(\sigma(V) < \infty) = 1$.

3. In Theorem 9 we give other conditions for validity of the equations (3.76) and (3.80).

THEOREM 9. *Suppose that the function* $g \in L(A^-,A^+)$ *and the conditions of Lemma 9 are satisfied. Then equaiton* (3.76) *holds. If, furthermore, for each point* $x \in E \setminus \Gamma_0$ *there is an open set* V *(with compact closure) such that* $\mathbf{P}_x(\sigma(V) < \infty) = 1$, *then* $s(x) \in \mathscr{D}_{\mathfrak{A}}(E \setminus \Gamma_0)$ *and equation* (3.80) *holds in the region* $C_0 = E \setminus \Gamma_0$.

PROOF. According to (3.63) and Lemma 5,

$$s(x) \leqq \tilde{\mathbf{E}}_x s(x_{\sigma_0}) \leqq \tilde{\mathbf{E}}_x s(x_{\sigma_0 \wedge \sigma(V)}) \leqq s(x).$$

As in Theorem 7, from this we obtain equation (3.76). Equation (3.80) can be derived as in Theorem 8.

§ 6. Optimal stopping of Markov processes and the generalized Stefan problem

1. Knowledge of the value $s(x)$ not only allows us to determine the magnitude of the optimal reward, but also (according to Theorems 4–6) makes it possible for us to construct optimal and ε-optimal Markov times. Under the assumptions of Theorems 8 and 9, the value $s(x)$ satisfies equation (3.80) in the region $C_0 = E \setminus \Gamma_0$ of "continuation of observations", and $s(x) = g(x)$ in the "termination" region Γ_0.

Thus the value is one of the solutions of the problem (3.80)–(3.81), which is characterized by the fact that not only the function $s(x)$, but also the region C_0 in which equation (3.81) operates, is unknown. In the theory of partial differential equations such problems are called *Stefan problems,* or free boundary problems.

As a rule, conditions (3.80)–(3.81) are not sufficient for finding the unknown function $s(x)$ and the region of "continuation of observations," C_0. Therefore we look for additional conditions which must be satisfied by the desired function $s(x)$.

Below we consider several cases of optimal stopping problems for Markov processes where we are able to find additional conditions satisfied by the function $s(x)$ on the boundary $\partial\Gamma_0$. These conditions, generally speaking, may also not be sufficient for finding the value $s(x)$. However, in the problems which we consider later in Chapter IV, these conditions allow us to determine the value $s(x)$ completely, and consequently also the "continuation region" C_0.

2. We assume that $X = (x_t, \mathscr{F}_t, \mathbf{P}_x)$, $t \in \boldsymbol{T}$, is a one-dimensional, nonterminating, continuous standard process in the phase space $(E, \mathscr{B})$, $E \subseteq R$. Let

$$s(x) = \sup_{\tau \in \mathfrak{M}} \mathbf{E}_x g(x_\tau), \qquad \Gamma_0 = \{x \in E: s(x) = g(x)\},$$

$C_0 = E \setminus \Gamma_0$, let $\partial\Gamma_0$ be the boundary of the set Γ_0, and let the point $y \in \partial\Gamma_0$. We assume that for sufficiently small $\rho > 0$ the set $V_\rho^-(y) = \{x: y - \rho < x < y\} \subseteq C_0$, and $V_\rho^+(y) = \{x: y + \rho > x \geqq y\} \subseteq \Gamma_0$. It is clear that $s(y) = g(y)$ if $y \in \partial\Gamma_0$, and $V_\rho^-(y) \cup V_\rho^+(y) = V_\rho(y)$, where $V_\rho(y) = \{x: |x - y| < \rho\}$.

Let $\sigma_\rho(y) = \inf\{t \geqq 0: x_t \in E \setminus V_\rho(y)\}$.

Below we shall make essential use of the following assumptions:

A_1: $g(y) = T_{\sigma_\rho(y)} g(y) + o(p), y \in \partial\Gamma_0$.

A_2. In some neighborhood $V_\rho^-(y) \cup \{y\}$ of the point $y \in \partial\Gamma_0$ there exist continuous "left" derivatives $d^-g(x)/dx$ and $d^-s(x)/dx$.

A_3. $\mathbf{P}_y(x_{\sigma_\rho(y)} = y - \rho) > 0$ for sufficiently small $\rho > 0$.

THEOREM 10. *Suppose $X = (x_t, \mathscr{F}_t, \mathbf{P}_x)$, $t \in \boldsymbol{T}$, is a one-dimensional, nonterminating, continuous standard process. Suppose the almost Borel function $g \in L(A^+)$ is nonnegative and $\mathscr{C}_0$-continuous. If assumptions A_1—A_3 are satisfied, then we have the "smooth pasting" condition at the point $y \in \partial\Gamma_0$:*

$$\frac{d^-s(x)}{dx} = \frac{d^-g(x)}{dx}\bigg|_{x=y\in\partial\Gamma_0}. \tag{3.82}$$

PROOF.[7] Let $f(x) = s(x) - g(x)$. Then

$$T_{\sigma_\rho(y)} f(y) = o(\rho). \tag{3.83}$$

Indeed, by Theorem 1 the value $s(x)$ is the s.e.m. of the function $g(x)$. Here, according to Lemma 5, $s(y) \geqq T_{\sigma_\rho(y)} s(y)$. Using condition A_1, we then find

$$g(y) = T_{\sigma_\rho(y)} g(y) + o(\rho) = s(y) \geqq T_{\sigma_\rho(y)} s(y),$$

whence

$$o(\rho) \geqq T_{\sigma_\rho(y)} s(y) - T_{\sigma_\rho(y)} g(y) = T_{\sigma_\rho(y)} f(y) \geqq 0$$

which proves (3.83).

In view of the continuity of the process X for sufficiently small $\rho > 0$,

[7] If $V_\rho^+(y) \in C_0$, then we replace condition (3.82) by the condition of equality of the "right" derivatives: $d^+s(x)/dx = d^+g(x)/dx|_{x=y\in\partial\Gamma_0}$. Assumption A_2 is also modified in the obvious fashion.

$$
\begin{aligned}
T_{\sigma_\rho(y)} f(y) &= \int_{(\sigma_\rho(y)<\infty)} f(x_{\sigma_\rho(y)}) d\mathbf{P}_y \\
&= \int_{(x_{\sigma_\rho(y)}=y-\rho)} f(x_{\sigma_\rho(y)}) d\mathbf{P}_y \\
&= \frac{d^- f(x)}{dx}\bigg|_{x=y} \cdot \int_{(x_{\sigma_\rho(y)}=y-\rho)} [x_{\sigma_\rho(y)} - y] d\mathbf{P}_y + R_1(\rho) \\
&= -\rho \frac{d^- f(x)}{dx}\bigg|_{x=y} \cdot \mathbf{P}_y\{x_{\sigma_\rho(y)} = y - \rho\} + R_1(\rho),
\end{aligned}
\tag{3.84}
$$

where $R_1(\rho) = o(\rho)$ by A_2.

According to A_3, $\mathbf{P}_y\{x_{\sigma_\rho(y)} = y - \rho\} > 0$ for sufficiently small $\rho > 0$. Hence by (3.83) and (3.84) we obtain the desired equality (3.82).

REMARK. The derivation of the "smooth pasting" conditions (3.82) presented above can be extended to the case in which the process X is n-dimensional (for more detail see **[34]**, **[35]**).

From Theorems 8 and 10 we have the following result.

THEOREM 11. *Suppose the conditions of Theorem* 8 *are satisfied, and conditions* A_1—A_3 *are satisfied for all points of the boundary* $\partial\Gamma_0$. *Then the value* $s(x)$ *is a solution of the generalized Stefan problem:*

$$\mathfrak{A}s(x) = 0, \qquad x \in C_0 = E \setminus \Gamma_0, \tag{3.85}$$

$$s(x) = g(x), \qquad \bar{x}, \in \Gamma_0, \tag{3.86}$$

$$\frac{d^-s(x)}{dx} = \frac{d^-g(x)}{dx}\bigg|_{x=y}, \quad y \in \partial\Gamma_0, \quad V_\rho^-(y) \subseteq C_0, \tag{3.87}$$

$$\frac{d^+s(x)}{dx} = \frac{d^+g(x)}{dx}\bigg|_{x=y}, \quad y \in \partial\Gamma_0, \quad V_\rho^+(y) \subseteq C_0. \tag{3.88}$$

3. In order to state the generalization of the result of Theorem 10 to the case of processes which are not necessarily continuous, we introduce the following assumptions:

B_1. $\mathbf{E}_y\sigma_\rho(y) = \beta(y)\rho + o(\rho)$ for $y \in \partial\Gamma_0$, where $0 \leqq \beta(y) < \infty$ and

$$\mathbf{P}_y\{| x_{\sigma_\rho(y)} - y | > \rho\} = 0 \left(\mathbf{E}_y\sigma_\rho(y)\right). \tag{3.89}$$

B_2. If $\beta(y) = 0$, then $g(y) = T_{\sigma_\rho(y)}g(y) + o(\rho)$, $y \in \partial\Gamma_0$; but if $\beta(y) > 0$, then $f(y) = s(y) - g(y) \in \mathscr{D}_{\mathfrak{A}}$ and the limit

$$\lim_{\rho \downarrow 0} \frac{1}{\mathbf{E}_y\sigma_\rho(y)} \int_{\{|x_{\sigma_\rho(y)} - y| > \rho\}} f(x_{\sigma_\rho(y)}) d\mathbf{P}_y \ \left(= \mathfrak{A}_2 f(y)\right) \tag{3.90}$$

exists.

B_3. In some neighborhood $V_\rho^-(y) \cup \{y\}$ of the point $t \in \partial\Gamma_0$ there exist continuous "left" derivatives $d^-g(x)/dx$ and $d^-s(x)/dx$.

B_4. $\mathbf{P}_y\{x_{\sigma_\rho(y)} = y - \rho\} = \alpha(y) + o(1)$ at the point $y \in \partial\Gamma_0$, where $0 \leqq \alpha(y) \leqq 1$.

THEOREM 12. *Suppose $X = (x_t, \mathscr{F}_t, \mathbf{P}_x)$, $t \in \boldsymbol{T}$, is a one-dimensional, nonterminating standard process. Suppose the almost Borel function $g = g(x)$ is bounded, nonnegative and $\mathscr{C}_0$-continuous. If conditions $\mathrm{B}_1 - \mathrm{B}_4$ are satisfied, then*

$$\alpha(x)\frac{d^- f(x)}{dx} = \beta(x)\mathfrak{A}_1 f(x)\Big|_{x=y\in\partial\Gamma_0}, \tag{3.91}$$

where $\mathfrak{A}_1 = \mathfrak{A} - \mathfrak{A}_2$.

PROOF. First suppose $\beta(y) = 0$. Then $T_{\sigma_\rho(y)} f(y) = o(\rho)$ by B_2, (cf. (3.83)). According to B_1 and B_4, taking into account the boundedness of the function $f(x)$, we obtain

$$\begin{aligned} T_{\sigma_\rho(y)} f(y) &= \int_{(\sigma_\rho(y)<\infty)} f(x_{\sigma_\rho(y)})d\mathbf{P}_y \\ &= \int_{(x_{\sigma_\rho(y)}=y-\rho)} f(x_{\sigma_\rho(y)})d\mathbf{P}_y + o(\rho) \\ &= -\frac{d^- f(x)}{dx}\Big|_{x=y} \cdot \int_{(x_{\sigma_\rho(y)}=y-\rho)} |\, x_{\sigma_\rho(y)} - y \,|\, d\mathbf{P}_y + o(\rho) \\ &= -\frac{d^- f(x)}{dx}\Big|_{x=y} \cdot \rho\alpha(y) + o(\rho). \end{aligned}$$

Hence by condition B_2

$$\alpha(x)\frac{d^- f(x)}{dx}\Big|_{x=y} = 0$$

which proves (3.91) in the case $\beta(y) = 0$.

Now suppose $\beta(y) > 0$. By assumption B_2, $f(y) \in \mathscr{D}_{\mathfrak{A}}$, whence

$$T_{\sigma_\rho(y)} f(y) = \mathfrak{A} f(y)\mathbf{E}_y\sigma_\rho(y) + o(\mathbf{E}_y\sigma_\rho(y)),$$

where $0 < \mathbf{E}_y\sigma_\rho(y) < \infty$. Therefore

$$T_{\sigma_\rho(y)} f(y) = \mathfrak{A} f(y)\beta(y)\rho + o(\rho). \tag{3.92}$$

Further,

$$\begin{aligned} T_{\sigma_\rho(y)} f(y) = &\int_{(|x_{\sigma_\rho(y)}-y|=\rho)} f(x_{\sigma_\rho(y)})d\mathbf{P}_y \\ &+ \int_{(|x_{\sigma_\rho(y)}-y|>\rho)} f(x_{\sigma_\rho(y)})\mathbf{P}_y, \end{aligned} \tag{3.93}$$

where, as in the case $\beta(y) = 0$,

$$\int_{(|x_{\sigma_\rho(y)}-y|=\rho)} f(x_{\sigma_\rho(y)})d\mathbf{P}_y = -\alpha(y)\frac{d^- f(x)}{dx}\Big|_{x=y} \cdot \rho + o(\rho). \tag{3.94}$$

According to (3.90)

$$\int_{(|x_{\sigma_\rho(y)}-y|>\rho)} f(x_{\sigma_\rho(y)})d\mathbf{P}_y = \mathbf{E}_y\sigma_\rho(y)\cdot\mathfrak{A}_2 f(y) + o(\mathbf{E}_y\sigma_\rho(y)) = \mathfrak{A}_2 f(y)\beta(y)\rho + o(\rho). \tag{3.95}$$

Then from (3.92) — (3.95) we obtain

$$\mathfrak{A} f(y)\beta(y)\rho + o(\rho) = -\alpha(y)\frac{d^- f(x)}{dx}\bigg|_{x=y}\cdot\rho + \beta(y)\mathfrak{A}_2 f(y)\rho + o(\rho),$$

which proves (3.91).

REMARK 1. If $\alpha(y) > 0$ and $\beta(y) = 0$, then condition (3.91) becomes the "smooth pasting" condition (3.82).

REMARK 2. A generalization of Theorem 12 to the case of n-dimensional processes is given in [**35**].

From Theorems 9 and 12 we immediately obtain the following result.

THEOREM 13. *Suppose the nonnegative function $g(x)$ is bounded and continuous, and the value $s(x)$ is lower semicontinuous. Moreover, suppose the conditions of Theorems 9 and 12 are satisfied and for all points of the boundary $\partial\Gamma_0$ conditions B_1 — B_4 are satisfied. Then the value $s(x)$ is a solution of the generalized Stefan problem*:

$$\mathfrak{A}s(x) = 0, \qquad x \in C_0 = E\setminus\Gamma_0, \tag{3.96}$$

$$s(x) = g(x), \qquad x \in \Gamma_0, \tag{3.97}$$

$$\alpha(y)\frac{d^- f(x)}{dx}\bigg|_{x=y} = \beta(y)\mathfrak{A}_1 f(y), \quad y\in\partial\Gamma_0, \quad V_\rho^-(y)\subseteq C_0, \tag{3.98}$$

$$\alpha(y)\frac{d^+ f(x)}{dx}\bigg|_{x=y} = \beta(y)\mathfrak{A}_1 f(y), \quad y\in\partial\Gamma_0, \quad V_\rho^+(y)\subseteq C_0, \tag{3.99}$$

where $f(x) = s(x) - g(x)$.

CHAPTER IV

SOME APPLICATIONS TO PROBLEMS IN MATHEMATICAL STATISTICS

§ 1. Sequential testing of two simple hypotheses. Bayesian formulation

1. In general terms the problem of testing two statistical hypotheses can be formulated as follows.

On some measurable space $(\Omega,\mathscr{F})$ we are given two probability measures $\mathbf{P}_0,\mathbf{P}_1$ and a sequence of random variables $\xi_1(\omega)$, $\xi_2(\omega),\cdots$ whose joint probability distribution is $\mathbf{P}_\theta$, where the parameter θ, taking on the two values 0 and 1, is unknown. The problem of interest to us is to determine with the minimum "loss" the true value of the parameter θ from the observations $\xi_1(\omega)$, $\xi_2(\omega),\cdots$.

Below we consider the case in which the variables $\xi_1(\omega)$, $\xi_2(\omega),\cdots$ form a sequence of independent identically distributed random variables (relative to each of the measures $\mathbf{P}_i$, $i = 0,1$) whose one-dimensional probability density (with respect to some measure μ) is $p_i(x)$.[1)]

Depending on the nature of the assumptions on the structure of the unknown parameter, we will examine the following two formulations (Bayesian and variational) of this problem.

Bayesian formulation. Suppose that on the measurable space $(\Omega,\mathscr{F})$ we are given a family of probability measures $\{\mathbf{P}_\pi, 0 \leqq \pi \leqq 1\}$ such that $\mathbf{P}_\pi = \pi\mathbf{P}_1 + (1-\pi)\,\mathbf{P}_0$. We assume that the unknown parameter $\theta = \theta(\omega)$ is a random variable taking on two values 1 and 0 with probabilities $\mathbf{P}_\pi\big(\theta(\omega) = 1\big) = \pi$ and $\mathbf{P}_\pi\big(\theta(\omega) = 0\big) = 1 - \pi$.

Further, let $\xi_1(\omega)$, $\xi_2(\omega),\cdots$ be a sequence of independent random variables under each of the measures $\mathbf{P}_0$ and $\mathbf{P}_1$. Let

$$\mathscr{F}_n = \sigma\{(\omega,\pi)\colon \pi, \xi_0(\omega),\cdots, \xi_n(\omega)\},$$

where $\pi \in [0,1]$, $\xi_0(\omega) \equiv 0$ and $n \in \boldsymbol{N}$; let $\mathfrak{M} = \{\tau\}$ be the class of Markov stopping

---

1) The last assumption is not a restriction, since we can always take, for example, $\mu = \frac{1}{2}(\mathbf{P}_0 + \mathbf{P}_1)$, where the measures $\mathbf{P}_i\{(-\infty, x]\} = \mathbf{P}_i\{\omega\colon \xi_1(\omega) \leqq x\}$.

times $\tau = \tau(\omega,\pi)$ (relative to $F = \{\mathscr{F}_n\}$); let $\mathfrak{D}_\tau = \{d\}$ be the class of $\mathscr{F}_\tau$-measurable functions $d = d(\omega,\pi)$ taking on the two values 1 and 0.

DEFINITION 1. We call the pair $\delta = (\tau,d)$, where $\tau \in \mathfrak{M}$ and $d \in \mathfrak{D}_\tau$, a *decision function* or a *decision rule,* and let $\Delta = \{\delta\}$ denote the class of all decision rules.

With regard to the given problem of testing the two hypotheses H_0: $\theta = 0$ and H_1: $\theta = 1$, the meaning of these decision rules is explained as follows.

The function $\tau = \tau(\omega,\pi)$ determines the time of termination of the process of observation, and the function $d = d(\omega,\pi)$ shows which of the hypotheses (on the basis of knowledge of the values $\pi, \xi_1(\omega), \xi_2(\omega),\cdots$) should be accepted. In the case $d(\omega, \pi) = 1$ we accept the hypothesis H_1, and if $d(\omega, \pi) = 0$ we accept H_0. In this interpretation $d = d(\omega, \pi)$ is called the *terminal decision rule.*

Let

$$\alpha(\pi, \delta) = \mathbf{P}_1\{\omega : d(\omega, \pi) = 0\}, \qquad \beta(\pi, \delta) = \mathbf{P}_0\{\omega : d(\omega, \pi) = 1\}$$

denote the error probabilities (of first and second kind) corresponding to the rule δ.

The average risk, or the loss caused by the choice of the decision rule $\delta = (\tau, d)$ is measured by the variable

$$\begin{aligned}\rho(\pi, \delta) = \pi[c\mathbf{E}_1\tau + a\mathbf{P}_1(\omega : d(\omega, \pi) = 0)] \\ + (1 - \pi)[c\mathbf{E}_0\tau + b\mathbf{P}_0(\omega : d(\omega, \pi) = 1)],\end{aligned} \tag{4.1}$$

where a, b, c are nonnegative constants.

DEFINITION 2. For each π, $0 \leqq \pi \leqq 1$, the decision rule $\delta^* = (\tau^*, d^*) \in \Delta$ is said to be *π-Bayesian* if

$$\rho(\pi, \delta^*) = \inf_{\sigma \in \Delta} \rho(\pi, \delta). \tag{4.2}$$

DEFINITION 3. The decision rule $\delta^* = (\tau^*,d^*) \in \Delta$ is said to be *Bayesian* (relative to the family $\{\mathbf{P}_\pi, 0 \leqq \pi \leqq 1\}$) if δ^* is a π-Bayesian rule for all $0 \leqq \pi \leqq 1$.

The problem of testing hypotheses in a Bayesian formulation consists of finding π-Bayesian and Bayesian rules. It will be shown further that the problem of finding these rules can be reduced to the solution of a special optimal stopping problem for some Markov process. From the solution of the latter problem it will follow that in the original problem of testing hypotheses there exist not only π-Bayesian but also Bayesian rules.

Variational formulation. In this formulation we make no probabilistic assumptions on the unknown parameter θ.

Suppose that on the measurable space $(\Omega,\mathscr{F})$ we have two probability measures $\mathbf{P}_0$ and $\mathbf{P}_1$ and a sequence of independent (under each of the measures $\mathbf{P}_i$, $i = 0,1$) random variables $\xi_1 = \xi_1(\omega)$, $\xi_2 = \xi_2(\omega),\cdots$. Let

$$\mathscr{F}_0^\xi = \{\phi,\Omega\}, \quad \mathscr{F}_n^\xi = \sigma\{\omega : \xi_1,\cdots, \xi_n\}, \qquad n \geqq 1,$$

let $\mathfrak{M}^\xi = \{\tau\}$ denote the class of stopping times $\tau = \tau(\omega)$ (relative to $F^\xi = \{\mathscr{F}_n^\xi\}$, $n \in N$), and let $\mathfrak{D}_\tau^\xi = \{d\}$ be the collection of $\mathscr{F}_\tau^\xi$-measurable functions $d = d(\omega)$ taking on the two values 0 and 1. Further let $\Delta^\xi(\alpha,\beta)$ be the class of decision rules $\delta = (\tau,d)$ such that $\mathbf{E}_0\tau < \infty$, $\mathbf{E}_1\tau < \infty$ and the error probabilities

$$\alpha(\delta) = \mathbf{P}_1\{\omega: d(\omega) = 0\} \leqq \alpha, \qquad \beta(\delta) = \mathbf{P}_0\{\omega: d(\omega) = 1\} \leqq \beta,$$

where the nonnegative numbers α and β satisfy the condition $\alpha + \beta < 1$.

Following A. Wald, we can formulate the variational problem of testing two simple hypotheses as follows [72].

Suppose we are given two nonnegative numbers α and β, $\alpha + \beta < 1$. We wish to find a rule $\tilde{\delta} = (\tilde{\tau},\tilde{d})$ in the class $\Delta^\xi(\alpha,\beta)$ such that simultaneously

$$\mathbf{E}_0\tilde{\tau} \leqq \mathbf{E}_0\tau, \qquad \mathbf{E}_1\tilde{\tau} \leqq \mathbf{E}_1\tau \tag{4.3}$$

for all $\delta = (\tau,d) \in \Delta^\xi(\alpha,\beta)$.

Wald's great success was to conjecture the existence of such a decision rule. In Theorem 4 we give a precise formulation of the conditions under which (4.3) holds, and the form of the decision rule $\tilde{\delta}$.

2. We proceed to a systematic investigation of the problem of testing hypotheses in the Bayesian formulation. First we shall show that finding a Bayesian rule can be reduced to solving an optimal stopping problem for some Markov process.

Suppose

$$\pi_n(\omega,\pi) = \mathbf{P}_\pi\{\theta(\omega) = 1 \mid \mathscr{F}_n^\xi\}, \qquad n \geqq 1,$$

is the posterior probability of the hypothesis H_1: $\theta(\omega) = 1$. We let $\pi_0(\omega,\pi) = \mathbf{P}_\pi\{\theta(\omega) = 1 \mid \mathscr{F}_0^\xi\}$, where $\mathscr{F}_0^\xi = \{\phi,\Omega\}$. Obviously $\mathbf{P}_\pi\{\pi_0(\omega,\pi) = \pi\} = 1$.

By Bayes' formula it is easy to find that with $\mathbf{P}_\pi$-probability one for all $n \geqq 1$.

$$\pi_n(\omega,\pi) = \frac{\pi p_1(\xi_1(\omega)) \ldots p_1(\xi_n(\omega))}{\pi p_1(\xi_1(\omega)) \ldots p_1(\xi_n(\omega)) + (1-\pi)p_0(\xi_1(\omega)) \ldots p_0(\xi_n(\omega))}. \tag{4.4}$$

Let

$$\varphi_n(\omega) = \frac{p_1(\xi_1(\omega)) \ldots p_1(\xi_n(\omega))}{p_0(\xi_1(\omega)) \ldots p_0(\xi_n(\omega))}. \tag{4.5}$$

Then for all $\pi < 1$

$$\pi_n(\omega) = \frac{\dfrac{\pi}{1-\pi}\varphi_n(\omega)}{1 + \dfrac{\pi}{1-\pi}\varphi_n(\omega)} \qquad (\mathbf{P}_\pi\text{-a.s.}). \tag{4.6}$$

It is easy to see that for all $n \geqq 0$

$$\varphi_{n+1}(\omega) = \varphi_n(\omega)\frac{p_1(\xi_{n+1}(\omega))}{p_0(\xi_{n+1}(\omega))} \qquad (\mathbf{P}_\pi\text{-a.s.}), \tag{4.7}$$

if we assume $\varphi_0(\omega) \equiv 1$, and

$$\pi_{n+1}(\omega) = \frac{\dfrac{\pi_n(\omega)}{1-\pi_n(\omega)} \cdot \dfrac{p_1(\xi_{n+1}(\omega))}{p_0(\xi_{n+1}(\omega))}}{1 + \dfrac{\pi_n(\omega)}{1-\pi_n(\omega)} \cdot \dfrac{p_1(\xi_{n+1}(\omega))}{p_0(\xi_{n+1}(\omega))}} \qquad (\mathbf{P}_\pi\text{-a,.s.}) \tag{4.8}$$

if $\mathbf{P}_\pi(\pi_n = 1) = 0$.

LEMMA 1. *For each* π, $0 \leqq \pi \leqq 1$, *the elements*

$$\Pi_\pi = \{\pi_n(\omega, \pi), \mathscr{F}_n^\xi, \mathbf{P}_\pi\}, \qquad n \in \boldsymbol{N},$$

form a Markov random function,[2] *i.e.* $\pi_n(\omega,\pi)$ *are* $\mathscr{F}_n^\xi$*-measurable functions and with* $\mathbf{P}_\pi$*-probability one*

$$\mathbf{P}_\pi\{\pi_{n+1}(\omega, \pi) \in B \mid \mathscr{F}_n^\xi\} = \mathbf{P}_\pi\{\pi_{n+1}(\omega, \pi) \in B \mid \pi_n\} \tag{4.9}$$

for every Borel set B *on* $[0,1]$.

PROOF. The $\mathscr{F}_n^\xi$-measurability of the functions $\pi_n(\omega,\pi)$ is obvious. To prove formula (4.9) it suffices to make use of (4.8), noting only that $\mathbf{P}_\pi$-a.s. for every Borel set $A \in R^1$

$$\mathbf{P}_\pi\{\xi_{n+1} \in A \mid \mathscr{F}_n^\xi\} = \pi_n(\omega,\pi)\mathbf{P}_1\{\xi_{n+1} \in A\} + (1 - \pi_n(\omega,\pi))\mathbf{P}_0\{\xi_{n+1} \in A\}.$$

REMARK 1. It can be proved analogously that the elements $\Phi_\pi = \{\varphi_n(\omega), \mathscr{F}_n^\xi, \mathbf{P}_\pi\}$, $n \in \boldsymbol{N}$, also form a Markov random function for each π, $0 \leqq \pi \leqq 1$. It is valuable to note that for given π, $0 < \pi < 1$, the statistics[3] $\pi_n(\omega,\pi)$ and $\varphi_n(\omega)$ are equivalent in the sense that from $\pi_n(\omega, \pi)$ we can uniquely establish the values of $\varphi_n(\omega)$, and vice versa.

REMARK 2. The statistics π_n, $n \in \boldsymbol{N}$ (as well as φ_n, $n \in \boldsymbol{N}$), have the important property that for every $n \in \boldsymbol{N}$ the variable π_{n+1} is determined by the values π_n and ξ_{n+1}. It is therefore natural to introduce

DEFINITION 4. Suppose that on the probability space $(\Omega,\mathscr{F},\mathbf{P})$ we are given a random sequence $\{x_n(\omega)\}$, $n \in \boldsymbol{N}$, with values in the measurable space $(E_x,\mathscr{B}_x)$. The collection of statistics $\{\gamma_n(\omega)\}$, $n \in \boldsymbol{N}$, with values in $(E_\gamma, \mathscr{B}_\gamma)$ is called a *transitive sequence* of statistics (relative to $F = \{\mathscr{F}_n^x\}$, $\mathscr{F}_n^x = \sigma\{\omega : x_0(\omega),\cdots,x_n(\omega)\}$, $n \in \boldsymbol{N}$) if for all $n \in \boldsymbol{N}$ the statistics $\gamma_n = \gamma_n(\omega)$ are $\mathscr{F}_n^x / \mathscr{B}_\gamma$-measurable and there exists

---

[2] We are using the terminology of **[25]**, Chapter 3, § 1.

[3] By a *statistic* we mean any (measurable) function of the results of observations.

a $\mathscr{B}_\gamma \times \mathscr{B}_x / \mathscr{B}_\gamma$-measurable function $\varphi_{n+1}(\gamma,x)$ such that with probability one

$$\gamma_{n+1}(\omega) = \varphi_{n+1}(\gamma_n(\omega), x_{n+1}(\omega)). \tag{4.10}$$

The result stated in Lemma 1 is a special case of the following general assertion.

LEMMA 2. *Suppose* $\{\gamma_n(\omega)\}, n \in \boldsymbol{N}$, *is a transitive sequence of statistics (relative to* $F = \{\mathscr{F}_n^x\}, n \in \boldsymbol{N}$) *and for each* $n \in \boldsymbol{N}$ *with probability one*

$$\mathbf{P}\{x_{n+1} \in B \mid \mathscr{F}_n^x\} = \mathbf{P}\{x_{n+1} \in B \mid \gamma_n\} \tag{4.11}$$

for all $B \in \mathscr{B}_x$. *Then the elements* $\{\gamma_n(\omega), \mathscr{F}_n^x, \mathbf{P}\}$, $n \in \boldsymbol{N}$, *form a Markov random function.*

PROOF. We need show only that for every bounded $\mathscr{B}_\gamma$-measurable function $f = f(\gamma)$

$$\mathbf{E}\{f(\gamma_{n+1}) \mid \mathscr{F}_n^x\} = \mathbf{E}\{f(\gamma_{n+1}) \mid \gamma_n\} \qquad (\mathbf{P}\text{-a.s.}). \tag{4.12}$$

By (4.10), $f(\gamma_{n+1}) = F_{n+1}(\gamma_n, x_{n+1})$, where $F_{n+1}(\gamma,x)$ is a bounded, $\mathscr{B}_\gamma \times \mathscr{B}_x$-measurable function. If $F_{n+1}(\gamma,x) = F_{n+1}^N(\gamma,x)$, where

$$F_{n+1}^N(\gamma,x) = \sum_{k=1}^N g_{k,n+1}(\gamma)\chi_{B_k}(x), \tag{4.13}$$

then from (4.11) we immediately obtain the ($\mathbf{P}$-a.s.) equality

$$\mathbf{E}\{F_{n+1}^N(\gamma_n, x_{n+1}) \mid \mathscr{F}_n^x\} = \mathbf{E}\{F_{n+1}^N(\gamma_n, x_{n+1}) \mid \gamma_n\}, \tag{4.14}$$

which proves (4.12) for functions of the form (4.13).

For the proof of (4.12) in the general case we have to construct a sequence of functions $F_{n+1}^N(\gamma,x)$, $N \geqq 1$ of the form (4.13), monotonely converging to $F_{n+1}(\gamma,x)$, and then pass to the limit in (4.14) as $N \to \infty$.

The theory of optimal stopping rules developed in the second chapter has the advantage that it makes it possible for us to find not only π-Bayesian rules for the particular values of π, but also Bayesian rules. As will become clear below, in applying the theory it turns out to be more convenient to work not with the family of Markov random functions $\{\Pi_x, 0 \leqq \pi \leqq 1\}$, but with its associated Markov process, which can be constructed as follows ([**25**], Chapter 3, § 1).

Suppose $\Omega' = \Omega \times [0,1]$ is the space of points $\omega' = (\omega,\pi)$, and $\mathscr{F}' = \mathscr{F} \times \mathscr{B}([0,11])$. For $\omega' = (\omega,\pi)$ and $n \geqq 0$, let

$$\pi_0'(\omega') = \pi, \qquad \pi_n'(\omega') = \pi_n(\omega,\pi), \qquad \xi_n'(\omega') = \xi_n(\omega),$$
$$\mathscr{F}_n' = \sigma\{\omega' : \pi_0'(\omega'), \xi_0'(\omega'), \ldots, \xi_n'(\omega')\},$$
$$\mathscr{F}_\infty' = \sigma\Big(\bigcup_{n \geqq 0} \mathscr{F}_n'\Big).$$

Further, let $A \in \mathscr{F}'_\infty$ and

$$A_\pi = \{\omega: (\omega,\pi) \in A\}, \qquad A^\omega = \{\pi: (\omega,\pi) \in A\}.$$

On $(\Omega',\mathscr{F}'_\infty)$ we define the measure

$$\mathbf{P}'_\pi(A) = \mathbf{P}_\pi(A_\pi)\cdot\chi_{A^\omega}(\pi), \qquad A \in \mathscr{F}'_\infty. \tag{4.15}$$

It is easy to verify that the elements $\Pi' = (\pi'_n,\mathscr{F}'_n,\mathbf{P}'_\pi)$, $\pi \in [0,1]$, $n \in \boldsymbol{N}$, form a Markov process. Since $\pi'_n(\omega,\pi) = \pi_n(\omega,\pi)$ and $\mathscr{F}'_n = \mathscr{F}_n$, we also use the notation $\Pi' = (\pi_n,\mathscr{F}_n,\mathbf{P}'_\pi)$ for the process Π'.

The possibility of reducing the problem of finding Bayesian rules to solving an optimal stopping problem for the Markov process $\Pi' = (\pi_n, \mathscr{F}_n, \mathbf{P}'_\pi)$ is based on the following lemma.

LEMMA 3. *Suppose $\delta = (\tau(\omega,\pi), d(\omega,\pi)) \in \Delta$ and the decision rule $\bar\delta = (\tau(\omega,\pi), d(\omega,\pi))$ is such that the time $\tau(\omega,\pi)$ is the same as in the rule δ, and*

$$d(\omega,\pi) = \begin{cases} 1, & a\pi_\tau \geqq b(1 - \pi_\tau), \\ 0, & a\pi_\tau < b(1 - \pi_\tau). \end{cases} \tag{4.16}$$

Then for all π, $0 \leqq \pi \leqq 1$,

$$\rho(\pi, \bar\delta) \leqq \rho(\pi, \delta) \tag{4.17}$$

and

$$\rho(\pi, \bar\delta) = \mathbf{E}_\pi\{c\tau + g(\pi_\tau)\}, \tag{4.18}$$

where $g(\pi) = \min(a\pi, b(1-\pi))$, and $\mathbf{E}_\pi$ is the average with respect to the measure $\mathbf{P}_\pi$.

PROOF. If $\pi = 0$ or 1, assertions (4.17) and (4.18) are obvious. Therefore we assume $0 < \pi < 1$.

For $0 < \pi < 1$ each of the measures $\mathbf{P}_i$, $i = 0, 1$, is absolutely continuous with respect to the measure $\mathbf{P}_\pi$. Let $(d\mathbf{P}_i / d\mathbf{P}_\pi)(\omega)$ be the Radon-Nikodým derivative of the measure $\mathbf{P}_i$ with respect to the measure $\mathbf{P}_\pi$ [67]. Then

$$\begin{aligned} \pi\mathbf{P}_1\{\omega: d(\omega, \pi) = 0\} &= \pi\mathbf{E}_1[1 - d(\omega, \pi)] \\ &= \pi\mathbf{E}_\pi[(1 - d(\omega, \pi))(d\mathbf{P}_1 / d\mathbf{P}_\pi)(\omega)] \\ &= \mathbf{E}_\pi\{(1 - d(\omega, \pi))\mathbf{E}_\pi[\pi d\mathbf{P}_1/d\mathbf{P}_\pi \mid \mathscr{F}^\xi_\tau]\} \\ &= \mathbf{E}_\pi\{(1 - d(\omega, \pi))\pi_\tau(\omega, \pi)\}. \end{aligned} \tag{4.19}$$

Here we made use of the equality

$$\mathbf{E}_\pi\left\{\pi\frac{d\mathbf{P}_1}{d\mathbf{P}_\pi}\middle| \mathscr{F}^\xi_\tau\right\} = \pi_\tau(\omega, \pi) \qquad (\mathbf{P}_\pi\text{-a.s.}), \tag{4.20}$$

which follows immediately from the fact that for all $n \geqq 0$ the conditional probability

$$\pi_n(\omega, \pi) = \pi(d\mathbf{P}_1/d\mathbf{P}_\pi)(\mathscr{F}_n^\xi)(\omega) \qquad (\mathbf{P}_\pi\text{-a.s.}),$$

where the Radon-Nikodým derivative $(d\mathbf{P}_1/d\mathbf{P}_\pi)(\mathscr{F}_n^\xi)(\omega)$ is an $\mathscr{F}_n^\xi$-measurable function such that for all $A \in \mathscr{F}_n^\xi$

$$\int_A \frac{d\mathbf{P}_1}{d\mathbf{P}_\pi}(\mathscr{F}_n^\xi)(\omega)d\mathbf{P}_\pi(\omega) = \mathbf{P}_1(A).$$

Analogously to (4.19), we can establish the equality

$$(1 - \pi)\mathbf{P}_0\{\omega: d(\omega, \pi) = 1\} = \mathbf{E}_\pi\{d(\omega, \pi)(1 - \pi_\tau(\omega, \pi))\}. \tag{4.21}$$

From (4.1), (4.19) and (4.21) we obtain

$$\begin{aligned}\rho(\pi, \delta) &= \mathbf{E}_\pi\{c\tau + a\pi_\tau[1 - d(\omega, \pi)] + b[1 - \pi_\tau]d(\omega, \pi)\} \\ &\geqq \mathbf{E}_\pi\{c\tau + \min[a\pi_\tau, b(1 - \pi_\tau)]\} = \rho(\pi, \bar{\delta}),\end{aligned}$$

which proves the lemma.

From this lemma it follows, in particular, that in seeking Bayesian rules we need consider only rules of the form $\bar{\delta} = (\tau,\bar{d})$, where the terminal decision $\bar{d} = \bar{d}(\omega,\pi)$ is given by formula (4.16), and that for all π, $0 \leqq \pi \leqq 1$, the "risk"

$$\rho(\pi) = \inf_{\delta \in \Delta} \rho(\pi, \delta) = \inf_{\tau \in \mathfrak{M}} \mathbf{E}_\pi\{c\tau(\omega, \pi) + g(\pi_\tau)\}. \tag{4.22}$$

Now we note that if the set $A \in \mathscr{F}'_\infty$ is such that $A^\omega = \{\pi: (\omega,\pi) \in A\} = [0,1]$, then $\mathbf{P}'_\pi(A) = \mathbf{P}_\pi(A_\pi)$ by (4.15). Therefore

$$\begin{aligned}\rho(\pi) &= \inf_{\tau \in \mathfrak{M}} \mathbf{E}_\pi\{c\tau(\omega, \pi) + g(\pi_\tau)\} \\ &= \inf_{\tau \in \mathfrak{M}} \mathbf{E}'_\pi\{c\tau(\omega') + g(\pi_\tau(\omega'))\}.\end{aligned} \tag{4.23}$$

We let $\mathfrak{M}^\pi = \{\tau\}$ denote the class of stopping times such that $\{\omega': \tau(\omega') = n\} \in \mathscr{F}_n^\pi$, where

$$\mathscr{F}_n^\pi = \sigma\{\omega': \pi_0(\omega'), \ldots, \pi_n(\omega')\}.$$

It is easy to see that $\mathfrak{M} \supseteq \mathfrak{M}^\pi$. Since the process $\Pi' = (\pi_n, \mathscr{F}'_n, \mathbf{P}'_\pi)$ is Markov, so is the process $\Pi'' = (\pi_n, \mathscr{F}_n^\pi, \mathbf{P}'_\pi)$, $n \in \boldsymbol{N}$. It then follows from the results of § 7.3 of Chapter II that the class of stopping times $\mathfrak{M}^\pi$ is sufficient, i.e.

$$\rho(\pi) = \inf_{\tau \in \mathfrak{M}} \mathbf{E}'_\pi\{c\tau + g(\pi_\tau)\} = \inf_{\tau \in \mathfrak{M}^\pi} \mathbf{E}'_\pi\{c\tau + g(\pi_\tau)\}. \tag{4.24}$$

According to Thoerem II. 16, the risk $\rho(\pi)$ satisfies[4] the recursion equation

---

[4] There is obviously no difficulty in reformulating the results obtained in Chapters II and III for the "value" to the case of a "risk."

$$\rho(\pi) = \min\{g(\pi), c + T\rho(\pi)\}, \tag{4.25}$$

where $T\rho(\pi) = \mathbf{E}'_\pi \rho(\pi_1)$. From the Corollary to this same theorem it follows that for $c > 0$ the time

$$\tau^*(\omega') = \inf\{n \geqq 0: \rho(\pi_n) = g(\pi_n)\} \tag{4.26}$$

is an optimal stopping time, i.e., for all π, $0 \leqq \pi \leqq 1$,

$$\rho(\pi) = \mathbf{E}'_\pi\{c\tau^* + g(\pi_{\tau^*})\}$$

and

$$\mathbf{P}'_\pi(\tau^* < \infty) = 1, \qquad 0 \leqq \pi \leqq 1.$$

Now we shall show that the function $\rho(\pi)$, $0 \leqq \pi \leqq 1$, is convex.
Let

$$Q_c g(\pi) = \min\{g(\pi), c + Tg(\pi)\}.$$

According to (4.4),

$$Tg(\pi) = \int_{-\infty}^{\infty} g\left(\frac{\pi p_1(x)}{\pi p_1(x) + (1-\pi)p_0(x)}\right)[\pi p_1(x) + (1-\pi)p_0(x)]d\mu(x).$$

Making use of the convexity of the function $g(\pi)$, we can now easily derive that the function $Tg(\pi)$ also has this property. Hence it follows that each of the functions $Q_c^N g(\pi)$ and $v(\pi) = \lim_{N\to\infty} Q_c^N g(\pi)$ is convex.

Since $0 \leqq Q_c^N g(\pi) \leqq ab/(a+b) < \infty$ for all $N \in \boldsymbol{N}$, we can show that the function $-v(x)$ is $(1,c)$-excessive (see § 8, Chapter II), by the same method used in the proof of Lemma II.4. It is also obvious that this function is a regular $(1,c)$-excessive majorant of $-g(x)$. Consequently

$$\rho(\pi) = v(\pi) = \lim_{N\to\infty} Q_c^N g(\pi),$$

which proves the convexity of the function $\rho(\pi)$.

From the convexity of the functions $\rho(\pi)$ and $T\rho(\pi)$ it follows that they are continuous on the interval $(0,1)$ ([**25**], Theorem 0.8). Therefore from (4.25) we obtain that the region of continuation of observations $C_0 = \{\pi: \rho(\pi) < g(\pi)\}$ has the form

$$C_0 = \{\pi: A^* < \pi < B^*\},$$

where $0 \leqq A^* \leqq B^* \leqq 1$.

It is easy to see that if $\rho(\pi) \equiv g(\pi)$ then $A^* = B^* = b/(a+b)$, and consequently the continuation region $C_0 = \phi$. If at even one point $\rho(\pi) < g(\pi)$, then $A^* < B^*$.

We have thus proved

THEOREM 1. *In the problem of sequential testing of two hypotheses the Bayesian rule* $\delta^* = (\tau^*, d^*)$ *exists and has the following form:*

$$\tau^*(\omega, \pi) = \inf\{n \geqq 0 : \pi_n(\omega, \pi) \notin (A^*, B^*)\}, \tag{4.27}$$

$$d^*(\omega, \pi) = \begin{cases} 1, & a\pi_{\tau^*} \geqq b(1 - \pi_{\tau^*}), \\ 0, & a\pi_{\tau^*} < b(1 - \pi_{\tau^*}), \end{cases} \tag{4.28}$$

where $\pi_0(\omega,\pi) = \pi$ *and* A^* *and* B^* *are constants,* $0 \leqq A^* \leqq B^* = 1$.

REMARK 1. If we pass from the statistics π_n to φ_n for given π, $0 < \pi < 1$, then the region of continuation of observations for the π-Bayesian decision rule will be of the form

$$\left\{\varphi : \frac{A^*}{1 - A^*} \cdot \frac{1 - \pi}{\pi} < \varphi < \frac{B^*}{1 - B^*} \cdot \frac{1 - \pi}{\pi}\right\}. \tag{4.29}$$

REMARK 2. The terminal decision $d^*(\omega,\pi)$ can also be written in the following form:

$$d^*(\omega, \pi) = \begin{cases} 1, & \pi_{\tau^*} \geqq B^*, \\ 0, & \pi_{\tau^*} \leqq A^*. \end{cases} \tag{4.30}$$

In order to describe fully the Bayesian rules $\delta^* = (\tau^*,d^*)$, we still have to determine the unknown constants A^* and B^* involved in (4.27). However, finding them is a very difficult task in the general case. In the next subsection we consider the problem of testing two simple hypotheses concerning the mean value of a Wiener process. In this case, working on the basis of the results presented in Chapter III, we are able to find a system of equations from which the constants A^* and B^* can be uniquely determined.

3. We assume that on the measurable space $(\Omega,\mathscr{F})$ we are given:

1) a family of probability measures $\{\mathbf{P}_\pi, 0 \leqq \pi \leqq 1\}$ such that $\mathbf{P}_\pi = \pi\mathbf{P}_1 + (1 - \pi)\mathbf{P}_0$;

2) a random variable $\theta = \theta(\omega)$ taking on the two values 1 and 0 with probabilities $\mathbf{P}_\pi(\theta(\omega) = 1) = \pi$ and $\mathbf{P}_\pi(\theta(\omega) = 0) = 1 - \pi$;

3) a standard Wiener process $\{w_t(\omega)\}$, $t \in \boldsymbol{T}$, which is independent of $\theta = \theta(\omega)$ (under each of the measures $\mathbf{P}_\pi$) and such that for all π, $0 \leqq \pi \leqq 1$,

$$w_0(\omega) \equiv 0, \quad \mathbf{E}_\pi \Delta w_t(\omega) = 0, \quad \mathbf{E}_\pi(\Delta w_t(\omega))^2 = \Delta t.$$

As above, we assume that the random variable $\theta = \theta(\omega)$ is unobservable, but we can observe the random process $\{\xi_t(\omega)\}$, $t \geqq 0$, where

$$\xi_t(\omega) = r \cdot \theta(\omega) \cdot t + \sigma w_t(\omega), \qquad \sigma > 0, \quad r \neq 0. \tag{4.31}$$

Analogously to the case of discrete time, we introduce the concepts of decision rules $\delta = (\tau(\omega,\pi), d(\omega,\pi))$, risk $\rho(\pi)$, and π-Bayesian rules.

We let

$$\pi_t(\omega,\pi) = \mathbf{P}_\pi\{\theta(\omega) = 1 \mid \mathscr{F}_t^\xi\}, \qquad \mathscr{F}_t^\xi = \sigma\{\omega: \xi_s(\omega),\, s \leqq t\},$$

and suppose that

$$\varphi_t(\omega) = (d\mathbf{P}_1/d\mathbf{P}_0)(\mathscr{F}_t^\xi)(\omega)$$

is the Radon-Nikodým derivative of the measure $\mathbf{P}_1$ with respect to the measure $\mathbf{P}_0$, i.e. it is an $\mathscr{F}_t^\xi$-measurable function such that for every set $A \in \mathscr{F}_t^\xi$

$$\int_A \frac{d\mathbf{P}_1}{d\mathbf{P}_0}(\mathscr{F}_t^\xi)(\omega)\,d\mathbf{P}_0(\omega) = \mathbf{P}_1(A).$$

It is known **[67]** that in this case

$$\varphi_t(\omega) = \exp\left\{\frac{r}{\sigma^2}\left\{\xi_t(\omega) - \frac{r}{2}t\right\}\right\} \qquad (\mathbf{P}_\pi\text{-a.s., } 0 \leqq \pi \leqq 1). \tag{4.32}$$

If $\pi = 1$, then $\mathbf{P}_1(\pi_t(\omega,1) = 1) = 1$ with $\mathbf{P}_1$-probability 1 for all $t \geqq 0$. In the case $\pi > 1$ the posterior probability

$$\pi_t(\omega,\pi) = \pi\frac{d\mathbf{P}_1}{d\mathbf{P}_\pi}(\mathscr{F}_t^\xi)(\omega) = \frac{(\pi/(1-\pi))\varphi_t(\omega)}{1+(\pi/(1-\pi))\varphi_t(\omega)}. \tag{4.33}$$

Now we note that

$$\varphi_{t+s}(\omega) = \varphi_t(\omega)\exp\left\{\frac{r}{\sigma^2}\left[\theta(\omega)\cdot s + (w_{t+s}(\omega) - w_s(\omega)) - \frac{r}{2}s\right]\right\}$$

and

$$\begin{aligned}\mathbf{P}_\pi\{\theta(\omega)\cdot s + (w_{t+s}(\omega) - w_s(\omega)) \leqq x \mid \mathscr{F}_t^\xi\} \\ = \pi_t(\omega,\pi)\mathbf{P}_\pi\{w_{t+s}(\omega) - w_s(\omega) \leqq x - s\} \\ + (1 - \pi_t(\omega,\pi))\mathbf{P}_\pi\{w_{t+s}(\omega) - w_s(\omega) \leqq x\}.\end{aligned}$$

Hence it can easily be derived (cf. Lemma 1) that for every π, $0 \leqq \pi \leqq 1$, the elements $\Pi_\pi = \{\pi_t(\omega,\pi), \mathscr{F}_t^\xi, \mathbf{P}_\pi\}$ and $\Phi_\pi = \{\varphi_t(\omega), \mathscr{F}_t^\xi, \mathbf{P}_\pi\}$ form Markov random functions. Moreover, using notation analogous to that introduced for discrete time, we can show that the process $\Pi' = (\pi_t(\omega'), \mathscr{F}_t, \mathbf{P}'_\pi)$, $t \in \boldsymbol{T}$, $0 \leqq \pi < 1$, is also Markov.

4. THEOREM 2. *A Bayesian rule $\delta^* = (\tau^*(\omega,\pi), d^*(\omega,\pi))$ in the problem of testing between hypotheses H_1: $\theta = 1$ and H_0: $\theta = 0$ from the results of observations on the process $\{\xi_t(\omega)\}$, $t \geqq 0$, exists and has the following form:*

$$\tau^*(\omega,\pi) = \inf\{t \geqq 0: \pi_t(\omega,\pi) \notin (A^*, B^*)\}, \tag{4.34}$$

$$d^*(\omega, \pi) = \begin{cases} 1, & \pi_{\tau^*}(\omega, \pi) \geqq B^*, \\ 0, & \pi_{\tau^*}(\omega, \pi) \leqq A^*. \end{cases} \tag{4.35}$$

The constants A^* and B^* can be uniquely determined from the system of transcendental equations

$$b + a = C\{\psi(A^*) - \psi(B^*)\}, \tag{4.36}$$

$$b(1 - B^*) = aA^* + (B^* - A^*)(a - C\psi(A^*)) + C(\Psi(B^*) - \Psi(A^*)), \tag{4.37}$$

where $C = c(r^2/2\sigma^2)^{-1}$,

$$\Psi(\pi) = (1 - 2\pi) \ln\frac{\pi}{1 - \pi}, \tag{4.38}$$

$$\psi(\pi) = \Psi'(\pi) = \left\{\frac{1 - \pi}{\pi} - \frac{\pi}{1 - \pi}\right\} + 2 \ln\frac{1 - \pi}{\pi}.$$

The risk is

$$\rho(\pi) = \begin{cases} g(\pi), & \pi \notin (A^*, B^*), \\ g(A^*) + (\pi - A^*)(a - C\psi(A^*)) + C(\Psi(\pi) - \Psi(A^*)), \\ \pi \in (A^*, B^*). \end{cases} \tag{4.39}$$

PROOF. Below we shall show that the function $f(\pi) = - C\Psi(\pi)$, where $\Psi(\pi)$ is defined in (4.38), is such that for each time $\tau = \tau(\omega)$ with $\mathbf{E}_\pi \tau < \infty$, $0 \leqq \pi \leqq 1$, we have the following relationship:

$$\mathbf{E}_\pi f(\pi_\tau) - f(\pi) = c\mathbf{E}_\pi \tau, \qquad 0 \leqq \pi \leqq 1.$$

Hence by the results of § 3.8 of Chapter III it follows that the given problem with a fee ($c(\pi) \equiv c$) can be reduced to solving a new problem where the fee is equal to zero. Making use of this fact, it is easy to show on the basis of the results of § 4 that in the given case a Bayesian rule exists and is given by (4.34) and (4.35). Therefore the basic difficulty is the proof of (4.36), (4.37) and (4.39).

We assume that the desired function $\rho(\pi)$ has continuous derivatives $d\rho(\pi)/d\pi$ in a neighborhood of the points A^* and B^*. Then, by Remark 1 to Theorem III.7, it follows from Theorem III.10 that the desired function $\rho(\pi)$ is a solution of the following Stefan problem:

$$\begin{aligned} &\mathfrak{A}\rho(\pi) = - c, \qquad \pi \in (A^*, B^*), \\ &\rho(\pi) = g(\pi), \qquad \pi \notin (A^*, B^*), \\ &\left.\frac{d\rho(\pi)}{d\pi}\right|_{\pi = A^*} = \left.\frac{dg(\pi)}{d\pi}\right|_{\pi = A^*}, \\ &\left.\frac{d\rho(\pi)}{d\pi}\right|_{\pi = B^*} = \left.\frac{dg(\pi)}{d\pi}\right|_{\pi = B^*}. \end{aligned} \tag{4.40}$$

In order to solve this problem we must make a detailed investigation of the structure of the processes $\{\pi_t\}$ and $\{\xi_t\}$, $t \in \boldsymbol{T}$.

5. To this end, we first establish the following lemma.

LEMMA 4. *Suppose that on the probability space* $(\Omega, \mathscr{F}, \mathbf{P})$ *we are given a standard Wiener process* $\{w_t\}$, $t \in \boldsymbol{T}$, *and a measurable real random process* $\{\theta_t\}$, $t \in \boldsymbol{T}$, *not depending on it and such that*

$$\mathbf{E}\,|\,\theta_t\,| < \infty, \quad \int_0^t \mathbf{E}\,|\,\theta_s\,|^2\, ds < \infty, \qquad t \geqq 0.$$

Suppose the process $\{\eta_t\}$, $t \in \boldsymbol{T}$, *allows the stochastic differential* ([**30**], *p*. 387)

$$d\eta_t = \theta_t dt + \sigma dw_t, \quad \eta_0(\omega) \equiv 0, \qquad \sigma > 0. \tag{4.41}$$

Then there is a standard Wiener process $\{w_t\}$, $t \in \boldsymbol{T}$, *such that*

$$d\eta_t = \bar{\theta}_t dt + \sigma d\bar{w}_t, \qquad \eta_0(\omega) \equiv 0, \tag{4.42}$$

where $\bar{\theta}_t = \mathbf{E}(\theta_t \,|\, \mathscr{F}_t^\eta)$, $\mathscr{F}_t^\eta = \sigma\{\omega : \eta_s(\omega),\ s \leqq t\}$.

PROOF. Let

$$\bar{\eta}_t = \eta_t - \int_0^t \mathbf{E}(\theta_s \,|\, \mathscr{F}_s^\eta) ds. \tag{4.43}$$

As can easily be verified, the process $\bar{Y} = (\bar{\eta}_t, \mathscr{F}_t^\eta, \mathbf{P})$, $t \in \boldsymbol{T}$, forms a martingale with continuous trajectories. It follows from (4.42) and (4.43) that

$$d\bar{\eta}_t = (\theta_t - \mathbf{E}(\theta_t \,|\, \mathscr{F}_t^\eta))dt + \sigma d\bar{w}_t. \tag{4.44}$$

Using Itô's formula for change of variables ([**30**], p. 387), we obtain from (4.44) that the process $\{\bar{\eta}_t^2\}$, $t \geqq 0$, allows the stochastic differential

$$d\bar{\eta}_t^2 = 2\bar{\eta}_t\, d\bar{\eta}_t + \sigma^2 dt.$$

Hence for all $t \geqq s$

$$\bar{\eta}_t^2 - \bar{\eta}_s^2 = 2 \int_s^t \bar{\eta}_u [\theta_u - \mathbf{E}(\theta_u \,|\, \mathscr{F}_u^\eta)] du \tag{4.45}$$

$$+ 2\sigma \int_s^t \bar{\eta}_u dw_u + \sigma^2 (t - s).$$

By the conditions of the lemma, $\int_s^t \mathbf{E}\,|\,\bar{\eta}_u\,|^2 du < \infty$. From this it follows that $\mathbf{E} \int_s^t \bar{\eta}_u\, dw_u = 0$ (see Property 2′ on p. 381 of [**30**]). It is also clear that $\int_s^t \mathbf{E}\,|\,\theta_u\, \eta_u\,|\, du < \infty$. Therefore

$$\mathbf{E}\left\{\int_s^t \bar{\eta}_u[\theta_u - \mathbf{E}(\theta_u \mid \mathscr{F}_u^\eta)du \mid \mathscr{F}_s^\eta\right\} = 0 \quad (\mathbf{P}\text{-a.s.}),$$

$$\mathbf{E}\left\{\int_s^t \bar{\eta}_u \, dw_u \mid \mathscr{F}_s^\eta\right\} = 0 \quad (\mathbf{P}\text{-a.s.}), \tag{4.46}$$

$$\mathbf{E}\{(\bar{\eta}_t - \bar{\eta}_s)^2 \mid \mathscr{F}_s^\eta\} = \mathbf{E}\{\bar{\eta}_t^2 - \bar{\eta}_s^2 \mid \mathscr{F}_s^\eta\} = \sigma^2(t-s) \qquad (\mathbf{P}\text{-a.s.}).$$

But it is well known ([**22**], Theorem 11.9) that a martingale $\bar{Y} = (\bar{\eta}_t, \mathscr{F}_t^\eta, \mathbf{P})$, $t \in \boldsymbol{T}$, with continuous trajectories and satisfying the conditions $\mathbf{E}\bar{\eta}_t^2 < \infty$, $t \geqq 0$, and (4.46) is a Wiener process with $\mathbf{E}\Delta\bar{\eta}_t = 0$ and $\mathbf{E}(\Delta\bar{\eta}_t)^2 = \sigma^2\Delta t$. Therefore, setting $\bar{w}_t = \bar{\eta}_t/\sigma$, we obtain the desired representation (4.42).

We apply this lemma to the process $\{\xi_t\}$, $t \in \boldsymbol{T}$, which by (4.31) has the stochastic differential

$$d\xi_t = r\theta dt + \sigma dw_t, \qquad \xi_0(\omega) \equiv 0, \quad \sigma > 0. \tag{4.47}$$

Then for each π, $0 \leqq \pi \leqq 1$, there is a standard Wiener process $w_t(\omega, \tau)$ such that

$$d\xi_t(\omega) = r\pi_t(\omega,\pi)dt + \sigma d\bar{w}_t(\omega,\pi), \qquad \xi_0(\omega) \equiv 0, \tag{4.48}$$

where

$$\pi_t(\omega,\pi) = \mathbf{P}_\pi\{\theta(\omega) = 1 \mid \mathscr{F}_t^\xi\} = \mathbf{E}_\pi\{\theta(\omega) \mid \mathscr{F}_t^\xi\}.$$

The representation (4.48) allows us to prove the following important result.

LEMMA 5. *For each* π, $0 \leqq \pi \leqq 1$, *the Markov random function* $\Pi_\pi = \{\pi_t(\omega,\pi), \mathscr{F}_t^\xi, \mathbf{P}_\pi\}$, $t \in \boldsymbol{T}$, *is a diffusion process with differential*

$$d\pi_t(\omega,\pi) = \frac{r}{\sigma}\pi_t(\omega,\pi)(1 - \pi_t(\omega,\pi))d\bar{w}_t, \pi_0(\omega,\pi) = \pi, \tag{4.49}$$

where $\{\bar{w}_t\}$, $t \geqq 0$, *is the standard Wiener process involved in* (4.48).

PROOF. Stochastically differentiating the right sides of (4.32) and (4.33), we find $(\varphi_t = \varphi_t(\omega),\ \pi_t = \pi_t(\omega,\pi))$

$$d\varphi_t = \frac{r}{\sigma^2}\varphi_t d\xi_t, \qquad \varphi_0(\omega) \equiv 1, \tag{4.50}$$

$$d\pi_t = -\frac{r^2}{\sigma^2}\pi_t^2(1 - \pi_t)dt + \frac{r}{\sigma^2}\pi_t(1 - \pi_t)d\xi_t, \qquad \pi_0(\omega,\pi) = \pi. \tag{4.51}$$

Hence, taking account of (4.48), we obtain the desired representation (4.49), from which, in particular, follows the diffusion nature of the process Π_π.

6. We now proceed to solution of the Stefan problem (4.40). The proof that the desired function $\rho(\pi)$ is given by (4.39) and that the unknown constants A^* and B^* can be determined from equations (4.36) and (4.37) is divided into two stages.

In the first stage we show that in the class of twice continuously differentiable functions a solution of problem (4.40) exists and is unique. Then we establish that this solution coincides with the risk $\rho(\pi)$.

From Lemma 5 it can easily be derived that

$$\Pi' = \{\pi_t, \mathscr{F}_t, \mathbf{P}'_\pi\}, \qquad 0 \leqq \pi \leqq 1, \quad t \geqq 0,$$

is a Markov diffusion process. According to Theorem 5.7 of **[25]**, the restriction of the operator $\mathfrak{A}$, corresponding to the process Π', to twice continuously differentiable functions $f = f(\pi)$ coincides (see (4.49)) with the differential operator of second order

$$\mathfrak{D} = (r^2/2\sigma^2)[\pi(1-\pi)]^2\, d^2/d\pi^2.$$

We investigate the solution of problem (4.40) in the class of twice continuously differentiable functions $f = f(\pi)$. Since $\mathfrak{A}f(\pi) = \mathfrak{D}f(\pi)$, we have

$$(r^2/2\sigma^2)[\pi(1-\pi)]^2\, d^2f/d\pi^2 = -c, \qquad \pi \in (A, B), \tag{4.52}$$

$$f(\pi) = g(\pi), \qquad \pi \notin (A, B), \tag{4.53}$$

$$f'(A) = a, \qquad f'(B) = -b, \tag{4.54}$$

where the constants A and B are also unknown, and $0 \leqq A \leqq B \leqq 1$.

We fix some number $0 \leqq A \leqq b/(a+b)$. It is easy to see that the solution $f = f(\pi)$ of equation (4.52) in the region $\pi > A$ satisfying the conditions $f(A) = aA$ and $f'(A) = a$ is given by the formula

$$f(\pi) = g(A) + (\pi - A)\{g'(A) - C\psi(A)\} + C\{\Psi(\pi) - \Psi(A)\}, \tag{4.55}$$

where $C = c(r^2/2\sigma^2)^{-1}$, and $\psi(\pi)$ and $\Psi(\pi)$ are defined in (4.38). Using the boundary conditions at the point B ($f(B) = b(1-B)$, $f'(B) = -b$), we obtain the following system for finding the unknown constants A and B:

$$b + a = C\{\psi(A) - \psi(B)\}, \tag{4.56}$$

$$b(1-B) = aA + (B-A)\{a - C\psi(A)\} + C\{\Psi(B) - \Psi(A)\}. \tag{4.57}$$

We shall show that the unknown constants A and B ($0 \leqq A \leqq B \leqq 1$) can be determined uniquely from (4.56)–(4.57).

To this end we transform the system (4.56)–(4.57) to the following form:

$$(b - a) + [a + C\psi(B)] = -[a + C\psi(A)], \tag{4.58}$$

$$b + C[B\psi(B) - \Psi(B)] = C[A\psi(A) - \Psi(A)]. \tag{4.59}$$

Let $x = A/(1-A)$ and $y = B/(1-B)$. Then from (4.58) and (4.59), taking account of (4.38), we obtain

$$(b - a) + \left[a + C\left(\frac{1}{y} - y - 2\ln y\right)\right] \tag{4.60}$$
$$= -\left[a + C\left(\frac{1}{x} - x - 2\ln x\right)\right],$$
$$b - C[y + \ln y] = - C[x + \ln x], \tag{4.61}$$

From (4.61) it follows that to each value $0 \leqq x < \infty$ corresponds a unique value $y = y_1(x) \geqq 0$, where $y_1(0) = 0$ and for all $x > 0$

$$\frac{dy_1(x)}{dx} = \frac{1 + 1/x}{1 + 1/y} > 0.$$

Consequently $y = y_1(x)$ is a nondecreasing function for $x \geqq 0$.

Analogously, from (4.60) it follows that to each x corresponds a unique value $y = y_2(x)$, where $y_2(0) = \infty$ and for all $x > 0$

$$\frac{dy_2(x)}{dx} = - \frac{1/x^2 + 1 + 2/x}{1/y^2 + 1 + 2/y} < 0.$$

Consequently there is a unique value $x^* > 0$ for which $y_1(x^*) = y_2(x^*)$. Hence it follows in an obvious way that the system of equations (4.58)–(4.59) also has a unique solution (A^*,B^*). Consequently in the class of twice continuously differentiable functions and constants $0 \leqq A \leqq B \leqq 1$ a solution of problem (4.52)–(4.53) exists and is unique. We let $f = f^*(\pi)$ denote the solution of this problem and show that $\rho(\pi) = f^*(\pi)$.

First we note that for $c > 0$

$$\rho(\pi) = \inf_{\tau \in \mathfrak{M}^\pi} \mathbf{E}'_\pi\{c\tau + g(\pi_\tau)\} = \inf_{\tau \in \mathfrak{R}^\pi} \mathbf{E}'_\pi\{c\tau + g(\pi_\tau)\},$$

where $\mathfrak{R}^\pi = \{\tau\}$ is the class of stopping times $\tau \in \mathfrak{M}^\pi$ for which $\mathbf{E}'_\pi\tau < \infty$ for all $0 \leqq \pi \leqq 1$ (see § 3, Chapter II).

Further, it is obvious that

$$\begin{aligned}\rho(\pi) &= \inf_{\tau \in \mathfrak{R}^\pi} \mathbf{E}'_\pi\{c\tau + g(\pi_\tau)\} \\ &= \inf_{\tau \in \mathfrak{R}^\pi} \mathbf{E}'_\pi\{[c\tau + f^*(\pi_\tau)] + [g(\pi_\tau) - f^*(\pi_\tau)]\} \\ &\geqq \inf_{\tau \in \mathfrak{R}^\pi} \mathbf{E}'_\tau\{c\tau + f^*(\pi_\tau)\} + \inf_{\tau \in \mathfrak{R}^\pi} \mathbf{E}'_\pi\{g(\pi_\tau) - f^*(\pi_\tau)\}.\end{aligned} \tag{4.62}$$

Suppose $A^* < \pi < B^*$. Since the weak infinitesimal operator $\tilde{\mathscr{A}} f^*(\pi) = \mathfrak{D} f^*(\pi) = - c$, from the Corollary to Theorem 5.1 in **[25]** it follows that for each Markov time τ such that $\mathbf{E}'_\pi\tau < \infty$ we have

$$\mathbf{E}'_\pi f^*(\pi_\tau) - f^*(\pi) = - c\mathbf{E}'_\pi\tau.$$

Consequently, for such times

$$\mathbf{E}'_\pi\{c\tau + f^*(\pi_\tau)\} = f^*(\pi). \tag{4.63}$$

Since $g(\pi) \geqq f^*(\pi)$ for all values π, $0 \leqq \pi \leqq 1$, we have

$$\inf_{\tau \in \mathfrak{R}^\pi} \mathbf{E}'_\pi\{g(\pi_\tau) - f^*(\pi_\tau)\} \geqq 0, \qquad \pi \in (A^*, B^*). \tag{4.64}$$

It is easy to show that for all π, $0 \leqq \pi \leqq 1$, the time

$$\tau^* = \inf\{t \geqq 0: \pi_t \notin (A^*, B^*)\}$$

has $\mathbf{E}'_\pi \tau^* < \infty$. Consequently $\tau^* \in \mathfrak{R}^\pi$ and $\mathbf{E}'_\pi\{g(\pi_{\tau^*}) - f^*(\pi_{\tau^*})\} = 0$. Hence by (4.64), (4.62) and (4.63) it follows that

$$\rho(\pi) = \inf_{\tau \in \mathfrak{R}^\pi} \mathbf{E}'_\pi\{c\tau + g(\pi_{\tau^*})\} \geqq \inf_{\tau \in \mathfrak{R}^\pi} \mathbf{E}'_\pi\{c\tau + f^*(\pi_\tau)\} = f^*(\pi).$$

But for the time τ^*

$$\mathbf{E}'_\pi\{c\tau^* + g(\pi_{\tau^*})\} = \mathbf{E}'_\pi\{c\tau^* + f^*(\pi_{\tau^*})\},$$

and therefore $\rho(\pi) = f^*(\pi)$ for all $A^* < \pi < B^*$.

Now suppose $\pi \notin (A^*, B^*)$. Then $f^*(\pi) = g(\pi)$, and since $\tilde{\mathscr{A}} g(\pi) = \mathfrak{D} g(\pi) = 0$, we again get, by the Corollary to Theorem 5.1 of **[25]**,

$$\mathbf{E}'_\pi f^*(\pi_\tau) = f(\pi), \quad \pi \notin (A^*, B^*),$$

for every $\tau \in \mathfrak{R}^\pi$. Thus for $\pi \notin (A^*, B^*)$

$$\begin{aligned} \rho(\pi) &\geqq \inf_{\tau \in \mathfrak{R}^\pi} \mathbf{E}'_\pi\{c\tau + f^*(\pi_\tau)\} \\ &\geqq \inf_{\tau \in \mathfrak{R}^\pi} \mathbf{E}'_\pi(c\tau) + f^*(\pi) = f^*(\pi) = g(\pi). \end{aligned}$$

But for the time τ^*

$$\mathbf{E}'_\pi\{c\tau^* + g(\pi_{\tau^*})\} = g(\pi), \qquad \pi \notin (A^*, B^*).$$

Thus for all $\pi \notin (A^*, B^*)$ also, the risk $\rho(\pi) = f^*(\pi)$.

To complete the proof of the theorem we need only establish that the function $f(\pi) = -c\Psi(\pi)$ satisfies the relationship

$$\mathbf{E}'_\pi f(\pi_\tau) - f(\pi) = c\mathbf{E}'_\pi \tau.$$

for all τ with $\mathbf{E}'_\pi \tau < \infty$, $0 \leqq \pi \leqq 1$.

By (4.49) and Itô's formula for change of variable

$$df(\pi_t) = c\,dt - \frac{2c}{(r/\sigma)}\left[(1 - 2\pi_t) + 2\pi_t(1 - \pi_t)\ln\frac{1 - \pi_t}{\pi_t}\right] d\bar{w}_t.$$

Hence, applying Theorem 1 from **[31]**, Chapter 2, § 4, we find

$$\mathbf{E}'_\pi \int_0^\tau \left[(1 - 2\pi_t) + 2\pi_t(1 - \pi_t)\ln\frac{1 - \pi_t}{\pi_t}\right] d\bar{w}_t = 0$$

and consequently

$$\mathbf{E}'_\pi f(\pi_\tau) - f(\pi)$$

$$= c\mathbf{E}'_\pi\tau - \frac{2c}{\left(\frac{r}{\sigma}\right)}\mathbf{E}'_\pi \int_0^\tau \left[(1-2\pi_t) + 2\pi_t(1-\pi_t)\ln\frac{1-\pi_t}{\pi_t}\right] d\bar{w}_t = c\mathbf{E}'_\pi\tau.$$

Theorem 2 is proved.

REMARK. In the symmetric case ($a = b$) it follows from (4.36) and (4.37) that $B^* = 1 - A$ and A^* can be determined as the (unique) root of the equation

$$\frac{a}{C} = \frac{1-A^*}{A^*} - \frac{A^*}{1-A^*} + 2\ln\frac{1-A^*}{A^*}.$$

§ 2. Sequential testing of two simple hypotheses. Variational formulation

1. We begin with the problem of testing hypotheses about the mean value of a Wiener process.

Suppose $\{w_t\}$, $t \geqq 0$, is a standard Wiener process given on the probability space $(\Omega,\mathscr{F},\mathbf{P})$. We assume that we are allowed to observe the process $\{\xi_t\}$, $t \geqq 0$, with differential

$$d\xi_t = r\theta dt + \sigma dw_t, \qquad \xi_0(\omega) \equiv 0, \quad r \neq 0, \quad \sigma > 0, \tag{4.65}$$

where the parameter θ, taking on the two values $\theta = 1$ (hypothesis H_1) and $\theta = 0$ (hypothesis H_0), is unknown. We let $\mathscr{F}_t^\xi = \sigma\{\omega: \xi_s,\ s \leqq t\}$, $\mathscr{F}_0^\xi = \{\phi,\Omega\}$, $\mathscr{F}_\infty^\xi = \sigma(\bigcup_{t\geqq 0}\mathscr{F}_t^\xi)$, and let $\mathbf{P}_i$ be the probability measures on $(\Omega,\mathscr{F}_\infty^\xi)$ induced by the process $\{\xi_t\}$, $t \geqq 0$, for $\theta = i$, $i = 0,1$.

Further suppose that $\mathfrak{M}^\xi = \{\tau\}$ is the collection of stopping times $\tau = \tau(\omega)$ (relative to $F^\xi = \{\mathscr{F}_t^\xi\}$, $t \geqq 0$) such that $\mathbf{E}_0\tau < \infty$ and $\mathbf{E}_1\tau < \infty$, where $\mathbf{E}_i$ denotes the mean with respect to the measure $\mathbf{P}_i$. Let $\mathfrak{D}_\tau^\xi = \{d\}$ denote the collection of $\mathscr{F}_\tau^\xi$-measurable functions $d = d(\omega)$ taking on the two values 0 and 1. The collection of decision rules $\delta = (\tau,d)$, where $\tau \in \mathfrak{M}^\xi$ and $d \in \mathfrak{D}_\tau^\xi$, such that

$$\alpha(\delta) = \mathbf{P}_1\{d(\omega) = 0\} \leqq \alpha,$$
$$\beta(\delta) = \mathbf{P}_0\{d(\omega) = 1\} \leqq \beta,$$

is denoted by $\Delta^\xi(\alpha,\beta)$.

THEOREM 3. *Suppose* $\alpha + \beta < 1$. *Then in the class* $\Delta^\xi(\alpha,\beta)$ *there exists a decision rule* $\bar{\delta} = (\bar{\tau},\bar{d})$ *such that for all* $\delta = (\tau,d) \in \Delta^\xi(\alpha,\beta)$

$$\mathbf{E}_1\bar{\tau} \leqq \mathbf{E}_1\tau, \quad \mathbf{E}_0\bar{\tau} \leqq \mathbf{E}_0\tau, \tag{4.66}$$

where

$$\bar{\tau}(\omega) = \inf\{t \geqq 0: \lambda_t \notin (\tilde{A}, \tilde{B})\},$$

$$\tilde{d}(\omega) = \begin{cases} 1, & \lambda_{\tilde{\tau}} \geqq \tilde{B}, \\ 0, & \lambda_{\tilde{\tau}} \leqq \tilde{A}, \end{cases} \tag{4.67}$$

and

$$\lambda_t(\omega) = \ln \varphi_t(\omega) = \frac{r}{\sigma^2}\left\{\xi_t(\omega) - \frac{r}{2}t\right\}, \qquad \lambda_0(\omega) \equiv 0, \tag{4.68}$$

$$\tilde{A} = \ln \frac{\alpha}{1-\beta}, \qquad \tilde{B} = \ln \frac{1-\alpha}{\beta},$$

$$\mathbf{E}_0\tilde{\tau}(\omega) = \frac{\omega(\beta, \alpha)}{\rho}, \qquad \mathbf{E}_1\tilde{\tau}(\omega) = \frac{\omega(\alpha, \beta)}{\rho}, \tag{4.69}$$

where

$$\omega(x, y) = (1-x)\ln\frac{1-x}{y} + x\ln\frac{x}{1-y} \tag{4.70}$$

and $\rho = r^2/2\sigma^2$.

For the proof of this theorem we need some auxiliary assertions, which are stated in the lemmas below.

2. Let

$$\lambda_t^x(\omega) = x + (r/\sigma^2)\left\{\xi_t(\omega) - \frac{r}{2}t\right\}, \qquad \lambda_t(\omega) = \lambda_t^0(\omega),$$

$$\tau_{A,B}^x(\omega) = \inf\{t \geqq 0 \colon \lambda_t^x(\omega) \notin (A, B)\},$$

and

$$\alpha(x) = \mathbf{P}_1(\lambda^x_{\tau^x_{A,B}} = A), \qquad \beta(x) = \mathbf{P}_0(\lambda^x_{\tau^x_{A,B}} = B),$$

where $A \leqq x \leqq B$.

LEMMA 6. *For all* $A \leqq x \leqq B$

$$\alpha(x) = \frac{e^A(e^{B-x} - 1)}{e^B - e^A}, \qquad \beta(x) = \frac{e^x - e^A}{e^B - e^A}. \tag{4.71}$$

PROOF. It is known ([**25**], Theorem 13.16) that $\alpha(x)$ is a solution of the differential equation

$$\alpha''(x) + \alpha'(x) = 0, \qquad A < x < B,$$

satisfying the boundary conditions $\alpha(A) = 1$ and $\alpha(B) = 0$. Analogously, $\beta(x)$ satisfies the equation

$$\beta''(x) - \beta'(x) = 0, \qquad A < x < B,$$

with conditions $\beta(B) = 1$ and $\beta(A) = 0$. Solving these equations, we obtain (4.71).

LEMMA 7. *Let* $m_i(x) = \mathbf{E}_i\tau^x_{A,B}(\omega)$, *where* $A \leqq x \leqq B$. *Then*

$$m_1(x) = \frac{1}{\rho}\left\{\frac{(e^B - e^{A+B-x})(B-A)}{e^B - e^A} + A - x\right\}, \tag{4.72}$$

$$m_0(x) = \frac{1}{\rho}\left\{\frac{(e^B - e^x)(B-A)}{e^B - e^A} - B + x\right\}. \tag{4.73}$$

PROOF. To derive (4.72) and (4.73), it is sufficient to note that the function $m_i(x)$ $(i = 0,1)$ is a solution of the equation

$$(r^2/2\sigma^2)m_i''(x) + (-1)^{1-i}(r^2/2\sigma^2)m_i'(x) = -1,$$

satisfying the boundary conditions $m_i(A) = m_i(B) = 0$ (see [25]).

LEMMA 8. *Suppose* $\{w_t\}$, $t \geqq 0$, *is a standard Wiener process*:

$$\mathbf{E}w_t = 0, \quad \mathbf{E}w_t^2 = t, \quad t \geqq 0, \quad w_0(\omega) \equiv 0. \tag{4.74}$$

Suppose further that $F = \{\mathscr{F}_t\}$, $t \geqq 0$, *is a nondecreasing system of* σ-*algebras* $\mathscr{F}_t \subseteq \mathscr{F}$ *such that for each t the random variables* w_t *are* $\mathscr{F}_t$-*measurable and for all* $h > 0$ *the process* $w_{t+h} - w_t$ *does not depend on any of the events of the* σ-*algebra* $\mathscr{F}_t$. *Then the Wald identities*

$$\mathbf{E}w_\tau = 0, \qquad \mathbf{E}w_\tau^2 = \mathbf{E}\tau, \tag{4.75}$$

are satisfied for every Markov time $\tau = \tau(\omega)$ (*relative to* $F = \{\mathscr{F}_t\}$, $t \geqq 0$) *such that* $\mathbf{E}\tau < \infty$.

PROOF. Since $\int_0^\infty \chi^2_{(\tau \geqq t)}(\omega)dt = \tau(\omega) < \infty$ with probability one, it follows that the stochastic integral of Itô ([30], Chapter VIII, § 2)

$$\int_0^\infty \chi_{(\tau \geqq t)}(\omega)dw_t(\omega)$$

is defined and

$$w_\tau(\omega) = \int_0^\infty \chi_{(\tau \geqq t)}(\omega)dw_t(\omega).$$

Since

$$\int_0^\infty \mathbf{E}\chi^2_{(\tau \geqq t)}(\omega)dt = \int_0^\infty \mathbf{P}(\tau \geqq t)dt = \mathbf{E}\tau < \infty,$$

then, by the known properties of Itô stochastic integrals ([30], property 2′ on p. 381)

$$\mathbf{E}w_\tau = \mathbf{E}\int_0^\infty \chi_{(\tau\geqq t)}(\omega)dw_t = 0,$$

$$\mathbf{E}w_\tau^2 = \mathbf{E}\left[\int_0^\infty \chi_{(\tau\geqq t)}(\omega)dw_t\right]^2 = \int \mathbf{E}\chi^2_{(\tau\geqq t)}(\omega)dt = \mathbf{E}\tau,$$

which proves the lemma.

REMARK 1. The assertions of the lemma can also be obtained directly from Theorem 1 of **[31]**, Chapter I, § 4.

REMARK 2. The assumption $\mathbf{E}\tau < \infty$ involved in the condition of Lemma 8 cannot, generally speaking, be weakened, as is shown by the following example.

Let $\tau = \inf\{t \geqq 0\colon w_t = 1\}$. From Lemmas 6 and 7 it is easy to derive that $\mathbf{E}\tau = \infty$ and $\mathbf{P}(\tau < \infty) = 1$, whence $1 = \mathbf{E}w_\tau^2 \neq \mathbf{E}\tau = \infty$.

REMARK 3. Let $\tau = \inf\{t \geqq 0\colon |\, w_t \,| = A\}$, where $A < \infty$. Then $\mathbf{E}\tau = A^2$. Indeed, let $\tau_N = \min(\tau,N)$. Then $\mathbf{E}w^2_{\tau_N} = \mathbf{E}\tau_N$ by Lemma 8, whence $\mathbf{E}\tau_N \leqq A^2$, and thus $\mathbf{E}\tau = \lim_{N\to\infty} \mathbf{E}\tau_N \leqq A^2 < \infty$. Again applying Lemma 8, we find that $\mathbf{E}w_\tau^2 = \mathbf{E}\tau$. Since $\mathbf{P}(\tau < \infty) = 1$, it follows that $\mathbf{E}\tau = \mathbf{E}w_\tau^2 = A^2$.

REMARK 4. Let $\tau = \inf\{t \geqq 0\colon |\, w_t \,| = a\sqrt{t+b}\}$, where $0 < n < \infty$ and $0 \leqq a < 1$. Then $\mathbf{E}\tau = a^2b/(1-a^2)$. For the proof we set $\tau_N = \min(\tau,N)$, $N \in \boldsymbol{N}$. Then $\mathbf{E}\tau_N = \mathbf{E}w^2_{\tau_N} \leqq a^2\mathbf{E}(\tau_N + b)$, i.e. $\mathbf{E}_{\tau_N} \leqq a^2b\,/\,(1-a^2)$. Consequently $\mathbf{E}\tau = \lim_{N\to\infty} \mathbf{E}\tau_N \leqq a^2b\,/(1-a^2) < \infty$. From the law of the iterated logarithm it follows that $\mathbf{P}(\tau < \infty) = 1$. Therefore

$$\begin{aligned}\mathbf{E}\tau = \mathbf{E}w_\tau^2 &= \int_{(\tau<\infty)} w_\tau^2 d\mathbf{P} \\ &= a^2 \int_{(\tau<\infty)} [\tau(\omega) + b]d\mathbf{P} \\ &= a^2[\mathbf{E}\tau + b].\end{aligned}$$

From this we obtain the desired formula for $\mathbf{E}\tau$.

LEMMA 9. *Suppose* $\{w_t\}$, $t \geqq 0$, *is a standard Wiener process and* $F = \{\mathcal{F}_t\}$, $t \geqq 0$, *is a system of* σ*-algebras satisfying the conditions of Lemma* 8. *If* $\tau = \tau(\omega)$ *is a Markov time* (*relative to* $F = \{\mathcal{F}_t\}$, $t \geqq 0$) *such that* $\mathbf{P}(\tau \leqq N) = 1$, $N < \infty$, *then for all* $-\infty < \lambda < \infty$

$$\mathbf{E}\exp\{\lambda w_\tau - (\lambda^2/2)\tau\} = 1. \tag{4.76}$$

PROOF. Let

$$\eta_t = \exp\{\lambda w_t - (\lambda^2/2)t\}.$$

Then it is easy to show that the process $(\eta_t,\mathcal{F}_t,\mathbf{P})$, $t \geqq 0$, is a martingale and $\mathbf{E}\eta_t = 1$. From known properties of martingales (**[22]**, Chapter VII, Theorem II.7) it follows that

$$\mathbf{E}\eta_\tau \leqq 3 \sup_{t \leqq N} \mathbf{E}\eta_t = 3 < \infty. \tag{4.77}$$

It is also obvious that $\lim_{t\to\infty} \int_{(\tau>t)} \eta_t d\mathbf{P} = 0$, since $\mathbf{P}(\tau \leqq N) = 1, N < \infty$. Applying Theorem I.6 to the martingale $(\eta_t, \mathscr{F}_t, \mathbf{P})$, we obtain equality (4.76).

REMARK. Equality (4.76) is none other than the well-known *fundamental identity* of sequential analysis for a Wiener process in the case of bounded Markov times.

The method of "truncation" applied below (which, however, we never used above) sometimes allows us to prove the validity of (4.76) for unbounded Markov times. Suppose, for example,

$$\tau = \inf\{t \geqq 0: |w_t| = a\sqrt{t+b}\}, \qquad 0 < b < \infty, \quad 0 \leqq a < 1.$$

We form the "truncated" time $\tau_N = \min(\tau, N)$. Then, by (4.76),

$$1 = \mathbf{E}\left[e^{\lambda w_\tau - (\lambda^2/2)\tau}; \tau \leqq N\right] + \mathbf{E}\left[e^{\lambda w_N - (\lambda^2/2)N}; \tau > N\right].$$

But

$$\begin{aligned} &\mathbf{E}\left[e^{\lambda w_N - (\lambda^2/2)N}; \tau > N\right] \\ &\quad = \int_{(\tau>N)} e^{\lambda w_N - (\lambda^2/2)N} d\mathbf{P} \leqq \int_{(\tau>N)} e^{\lambda a\sqrt{N+b} - \lambda^2 N/2} d\mathbf{P} \to 0, \qquad N \to \infty. \end{aligned}$$

Therefore the identity (4.76) is also satisfied for the Markov time under consideration.

3. PROOF OF THEOREM 3. First we shall show that for each rule $\delta = (\tau(\omega), d(\omega)) \in \Delta^\xi(\alpha,\beta)$ we have

$$\mathbf{E}_0\tau \geqq w(\beta,\alpha)/\rho, \quad \mathbf{E}_1\tau \geqq w(\alpha,\beta)/\rho, \tag{4.78}$$

where the function $\omega(x,y)$ is defined in (4.70).

By (4.32) and (4.75),

$$\begin{aligned} \mathbf{E}_1 \ln\varphi_\tau(\omega) &= \mathbf{E}_1 \ln\frac{d\mathbf{P}_1}{d\mathbf{P}_0}(\mathscr{F}_\tau^\xi)(\omega) = \mathbf{E}_1\left\{\frac{r}{\sigma^2}\xi_\tau - \frac{r^2}{2\sigma_2}\tau\right\} \\ &= \mathbf{E}_1\left\{\frac{r^2}{2\sigma^2}\tau + \frac{r}{\sigma}w_\tau\right\} = \frac{r^2}{2\sigma^2}\mathbf{E}_1\tau = \rho\mathbf{E}_1\tau. \end{aligned} \tag{4.79}$$

On the other hand[5]

[5] If $\mathbf{P}_1(d(\omega) = i) = 0$, then we set the product

$$\mathbf{P}_1(d(\omega) = i) \cdot \int_\Omega \ln\frac{d\mathbf{P}_0}{d\mathbf{P}_1} d\mathbf{P}_1(\omega | d(\omega) = i)$$

equal to zero.

$$
\begin{aligned}
\mathbf{E}_1 \ln \varphi_\tau(\omega) &= -\mathbf{E}_1 \ln \frac{d\mathbf{P}_0}{d\mathbf{P}_1}(\mathscr{F}_\tau^\xi)(\omega) \\
&= -\int_{\{\omega: d(\omega)=1\}} \ln \frac{d\mathbf{P}_0}{d\mathbf{P}_1} d\mathbf{P}_1 - \int_{\{\omega: d(\omega)=0\}} \ln \frac{d\mathbf{P}_0}{d\mathbf{P}_1} d\mathbf{P}_1 \\
&= -\mathbf{P}_1(d(\omega)=1) \int_\Omega \ln \frac{d\mathbf{P}_0}{d\mathbf{P}_1} d\mathbf{P}_1(\omega \mid d(\omega)=1) \\
&\quad -\mathbf{P}_1(d(\omega)=0) \int_\Omega \ln \frac{d\mathbf{P}_0}{d\mathbf{P}_1} d\mathbf{P}_1(\omega \mid d(\omega)=0) \\
&\geqq -\mathbf{P}_1(d(\omega)=1) \ln \int_\Omega \frac{d\mathbf{P}_0}{d\mathbf{P}_1} d\mathbf{P}_1(\omega \mid d(\omega)=1) \\
&\quad -\mathbf{P}_1(d(\omega)=0) \ln \int_\Omega \frac{d\mathbf{P}_0}{d\mathbf{P}_1} d\mathbf{P}_1(\omega \mid d(\omega)=0),
\end{aligned} \tag{4.80}
$$

where we made use of Jensen's inequality

$$
\ln \mathbf{E}\eta(\omega) \geqq \mathbf{E}\ln \eta(\omega),
$$

which holds for every nonnegative random variable $\eta(\omega)$.

Transforming the right side of (4.80), we find

$$
\begin{aligned}
&\mathbf{E}_1 \ln \varphi_\tau(\omega) \\
&\geqq -\mathbf{P}_1(d(\omega)=1)\ln\left[\frac{1}{\mathbf{P}_1(d(\omega)=1)} \int_{\{\pi: d(\omega)=1\}} \frac{d\mathbf{P}}{d\mathbf{P}_1} d\mathbf{P}_1\right] \\
&\quad -\mathbf{P}_1(d(\omega)=0)\ln\left[\frac{1}{\mathbf{P}_1(d(\omega)=0)} \int_{\{\omega: d(\omega)=0\}} \frac{d\mathbf{P}_0}{d\mathbf{P}_1} d\mathbf{P}_1\right] \\
&= -\mathbf{P}_1(d(\omega)=1)\ln \frac{\mathbf{P}_0(d(\omega)=1)}{\mathbf{P}_1(d(\omega)=1)} \\
&\quad -\mathbf{P}_1(d(\omega)=0)\ln \frac{\mathbf{P}_0(d(\omega)=0)}{\mathbf{P}_1(d(\omega)=0)} \geqq -(1-\alpha)\ln\frac{\beta}{1-\alpha} \\
&\quad -\alpha\ln\frac{1-\beta}{\alpha} = (1-\alpha)\ln\frac{1-\alpha}{\beta} + \alpha\ln\frac{\alpha}{1-\beta} = \omega(\alpha,\beta).
\end{aligned} \tag{4.81}
$$

Equating (4.79) and (4.81), we arrive at the second formula in (4.78).

The first inequality in (4.78) can be proved analogously.

Now we consider the decision rule $\tilde{\delta} = (\tilde{\tau},\tilde{d})$ defined in (4.67). According to Lemma 6,

$$
\mathbf{P}_1(\tilde{d}(\omega)=0) = \alpha(0) = e^{\tilde{A}}(e^{\tilde{B}}-1)/(e^{\tilde{B}}-e^{\tilde{A}}) = \alpha \tag{4.82}
$$

and

$$
\mathbf{P}_0(\tilde{d}(\omega)=1) = \beta(0) = (1-e^{\tilde{A}})/(e^{\tilde{B}}-e^{\tilde{A}}) = \beta. \tag{4.83}
$$

Further, by Lemma 7,

$$\mathbf{E}_1\tilde{\tau} = \frac{1}{\rho}\left\{\frac{(e^{\tilde{B}} - e^{\tilde{A}+\tilde{B}})(\tilde{B} - \tilde{A})}{e^{\tilde{B}} - e^{\tilde{A}}} + \tilde{A}\right\}$$

$$= \frac{1}{\rho}\,\frac{\tilde{B}e^{\tilde{B}}(1 - e^{\tilde{A}}) + Ae^{\tilde{A}}(e^{\tilde{B}} - 1)}{e^{\tilde{B}} - e^{\tilde{A}}} = \frac{1}{\rho}\,\omega(\alpha, \beta) \tag{4.84}$$

and

$$\mathbf{E}_0\tilde{\tau} = \frac{1}{\rho}\left\{\frac{(e^{\tilde{B}} - 1)(\tilde{B} - \tilde{A})}{e^{\tilde{B}} - e^{\tilde{A}}} - \tilde{B}\right\}$$

$$= \frac{1}{\rho}\,\frac{\tilde{B}(e^{\tilde{A}} - 1) + \tilde{A}(1 - e^{\tilde{B}})}{e^{\tilde{B}} - e^{\tilde{A}}} = \frac{1}{\rho}\,\omega(\beta, \alpha). \tag{4.85}$$

From (4.82)–(4.85) it follows that the decision rule $\tilde{\delta} = (\tilde{\tau},\tilde{d}) \in \Delta^{\xi}(\alpha,\beta)$, and from a comparison of the inequalities (4.78) with (4.84) and (4.85) it follows that for every decision rule $\delta = (\tau,d) \in \Delta^{\xi}(\alpha,\beta)$

$$\mathbf{E}_0\tau \geqq \mathbf{E}_0\tilde{\tau}, \qquad \mathbf{E}_1\tau \geqq \mathbf{E}_1\tilde{\tau}.$$

Thus $\tilde{\delta} = (\tilde{\tau},\tilde{d})$ is an optimal rule in the variational formulation.

4. In an example of the problem of testing two simple hypotheses in the case of a Wiener process, we compare the mean times $\mathbf{E}_1\tilde{\tau}$ and $\mathbf{E}_0\tilde{\tau}$, corresponding to the optimal decision rule $\tilde{\delta} = (\tilde{\tau},\tilde{d}) \in \Delta^{\xi}(\alpha,\beta)$, with the fixed time of observation $t(\alpha,\beta)$ required for testing the hypotheses H_1: $\theta = 1$ and H_0: $\theta = 0$, if we make use of the most powerful classical rule **[45]** for which the probabilities of errors of first and second kind do not exceed α and β respectively.

Suppose $\mathscr{F}_0^{\xi} = \{\phi,\Omega\}$ and $\mathscr{F}_t^{\xi} = \sigma\{\omega\colon \xi_s(\omega),\, s \leqq t\}$, $t > 0$. Let $\tau_t(\omega)$ denote an arbitrary $\mathscr{F}_0^{\xi}$-measurable function such that $\tau_t(\omega) \equiv t$, and let $d_t(\omega)$ be any $\mathscr{F}_t^{\xi}$-measurable function taking on the two values 1 and 0.

Every pair of functions $\delta_t = (\tau(\omega),d_t(\omega))$ determines a classical rule with length of observation time equal to $\tau_t(\omega) \equiv t$ and terminal decision $d_t(\omega)$. If $d_t(\omega) = 1$ then we accept the hypothesis H_1; in the case $d_t(\omega) = 0$ we accept the hypothesis H_0.

Suppose $\Delta_0^{\xi}(\alpha,\beta)$ is the collection of classical rules $\delta_t = (\tau_t(\omega), d_t(\omega))$, $t \geqq 0$, for which the error probabilities are

$$\mathbf{P}_1\{d_t(\omega) = 0\} \leqq \alpha, \qquad \mathbf{P}_0\{d_t(\omega) = 1\} \leqq \beta.$$

It is obvious that $\Delta^{\xi}(\alpha,\beta) \supseteq \Delta_0^{\xi}(\alpha,\beta)$.

According to the fundamental Neyman-Pearson lemma **[45]**, for the most powerful classical rule

$$\delta_{t(\alpha,\beta)} = (t(\alpha,\beta),\, d_{t(\alpha,\beta)}(\omega)) \in \Delta_0^{\xi}(\alpha,\beta)$$

the terminal decision $d_{t(\alpha,\beta)}(\omega)$ is defined by the formula

$$d_{t(\alpha,\beta)}(\omega) = \begin{cases} 1, & \lambda_{t(\alpha,\beta)}(\omega) \geqq h(\alpha,\beta), \\ 0, & \lambda_{t(\alpha,\beta)}(\omega) < h(\alpha,\beta), \end{cases} \tag{4.86}$$

where the duration of observations $t(\alpha,\beta)$ and the "threshhold" $h(\alpha,\beta)$ are chosen so that the rule $\delta_{t(\alpha,\beta)}$ belongs to $\Delta_0^\xi(\alpha,\beta)$.

We shall show that

$$t(\alpha,\beta) = (C_\alpha + C_\beta)^2/2\rho, \tag{4.87}$$

$$h(\alpha,\beta) = (C_\beta^2 - C_\alpha^2)/2, \tag{4.88}$$

where C_γ is the root of the equation

$$\frac{1}{\sqrt{2\pi}}\int_{C_\gamma}^{\infty} e^{-x^2/2}dx = \gamma, \qquad 0 \leqq \gamma \leqq 1.$$

Indeed, for every rule $\delta_t = (t,d_t(\omega))$ such that

$$d_t(\omega) = \begin{cases} 1, & \lambda_t(\omega) \geqq h, \\ 0, & \lambda_t(\omega) < h, \end{cases}$$

the probability

$$\begin{aligned} \mathbf{P}_0\{d_t(\omega) = 1\} &= \mathbf{P}_0\{\lambda_t(\omega) \geqq h\} \\ &= \mathbf{P}_0\left\{\frac{r}{\sigma^2}\left[\xi_t(\omega) - \frac{r}{2}t\right] \geqq h\right\} \mathbf{P}\left\{\frac{r}{\sigma}\,\omega_t \geqq h + \frac{r^2}{2\sigma^2}t\right\} \\ &= \mathbf{P}\left\{\frac{\omega_t}{\sqrt{t}} \geqq \frac{h+\rho t}{\frac{r}{\sigma}\sqrt{t}}\right\} = \Phi\left(\frac{h+\rho t}{\frac{r}{\sigma}\sqrt{t}}\right), \end{aligned} \tag{4.89}$$

where $\Phi(x) = (1/\sqrt{2\pi})\int_x^\infty e^{-x^2/2}dx$. In precisely the same way

$$\mathbf{P}_1\{d_t(\omega) = 0\} = 1 - \Phi\left(\frac{h - \rho t}{(r/\sigma)\sqrt{t}}\right). \tag{4.90}$$

Equating the right sides of (4.89) and (4.90) to β and α respectively, we obtain a system of two equations for $t = t(\alpha,\beta)$ and $h = h(\alpha,\beta)$:

$$\frac{h+\rho t}{(r/\sigma)\sqrt{t}} = C_\beta, \qquad \frac{h-\rho t}{(r/\sigma)\sqrt{t}} = -C_\alpha,$$

from which (4.87) and (4.88) follow immediately.

Thus for given α and β, $\alpha + \beta < 1$, by (4.69) and (4.87) we obtain

$$\mathbf{E}_0\tilde{\tau}/t(\alpha,\beta) = 2\omega(\beta,\alpha)/(C_\alpha + C_\beta)^2, \tag{4.91}$$

$$\mathbf{E}_1\tilde{\tau}/t(\alpha,\beta) = 2\omega(\beta,\alpha)/(C_\alpha + C_\beta)^2. \tag{4.92}$$

Numerical computations [1] show that for $\alpha,\beta \leqq 0.03$

$$\mathbf{E}_0\tilde{\tau} \leqq (17/30)t(\alpha,\beta), \qquad \mathbf{E}_1\tilde{\tau} \leqq (17/30)t(\alpha,\beta).$$

Moreover, if $\alpha = \beta$, then (see [1])

$$\lim_{\alpha \downarrow 0} \frac{\mathbf{E}_0 \tilde{\tau}}{t(\alpha, \alpha)} = \lim_{\alpha \downarrow 0} \frac{\mathbf{E}_1 \tilde{\tau}}{t(\alpha, \alpha)} = \frac{1}{4}.$$

5. Now we proceed to the problem of testing two simple hypotheses in the variational formulation for the case of discrete time.

The basic result (cf. Theorem 3) is stated as follows.

THEOREM 4. *Suppose the nonnegative numbers α and β are such that $\alpha + \beta < 1$, and there are numbers $\tilde{A}$ and $\tilde{B}$, $\tilde{A} < 0 < \tilde{B}$, having the property that for the rule $\tilde{\delta} = (\tilde{\tau}, \tilde{d})$, where*

$$\tilde{\tau}(\omega) = \inf\{n \geqq 0: \lambda_n(\omega) \notin (\tilde{A}, \tilde{B})\},$$
$$\tilde{d}(\omega) = \begin{cases} 1, & \lambda_{\tilde{\tau}}(\omega) \geqq \tilde{B}, \\ 0, & \lambda_{\tilde{\tau}}(\omega) \leqq \tilde{A}, \end{cases} \tag{4.93}$$

the error probabilities $\alpha(\tilde{\delta})$ and $\beta(\tilde{\delta})$ are identically equal to α and β respectively.

Then the rule[6] *$\tilde{\delta} = (\tilde{\tau}, \tilde{d})$ in the class $\Delta^{\xi}(\alpha, \beta)$ is optimal in the sense that for every $\delta = (\tau, d) \in \Delta^{\xi}(\alpha, \beta)$*

$$\mathbf{E}_0 \tilde{\tau} \leqq \mathbf{E}_0 \tau, \qquad \mathbf{E}_1 \tilde{\tau} \leqq \mathbf{E}_1 \tau.$$

The proof of the optimality of the rule $\tilde{\delta} = (\tilde{\tau}, \tilde{d})$ is essentially based on the properties of the Bayesian rule $\delta^* = (\tau^*, d^*)$ (see Theorem 1) given in [45] (Chapter 3, § 12), and will not be presented here.

REMARK 1. Above we assumed that $\alpha + \beta < 1$. The case where $\alpha + \beta \geqq 1$ is not of great interest, in view of the following.

Consider the randomized decision rule which consists of accepting the hypothesis H_0 with probability $1 - \alpha$ and H_1 with probability α, without observations. Moreover, suppose $(\tilde{\Omega}, \tilde{\mathscr{F}}, \tilde{\mathbf{P}})$ is an auxiliary probability space and $\eta = \eta(\tilde{\omega})$, $\tilde{\omega} \in \tilde{\Omega}$, is a random variable taking on the two values 0 and 1 with probabilities $1 - \alpha$ and α respectively. If $\eta = 0$, then we accept the hypothesis H_0. If $\eta = 1$, we accept the hypothesis H_1. For such a randomized decision rule the duration of observation is equal to zero and the error probabilities satisfy the given restrictions.

REMARK 2. Theorem 4 gives conditions for optimality of the rule $\tilde{\delta} = (\tilde{\tau}, \tilde{d})$ in the class of rules $\delta = (\tau, d) \in \Delta^{\xi}(\alpha, \beta)$ for which $\mathbf{E}_0 \tau < \infty$ and $\mathbf{E}_1 \tau < \infty$. In fact it can be shown [13] that the rule $\tilde{\delta}$ is also optimal in the wider class of rules $\delta = (\tau, d)$ for which $\mathbf{E}_0 \tau$ and $\mathbf{E}_1 \tau$ can also take on infinite values.

REMARK 3. It can happen that for given α and β and all threshholds A and B the error probabilities of first and second kind are not exactly equal to α and β. In this case Theorem 4 does not guarantee that among the rules $\delta_{(A,B)} = (\tau_{(A,B)}, d_{(A,B)}) \in \Delta^{\xi}(\alpha, \beta)$ such that

[6] If we pass from λ_t to the statistic $\varphi_t = e^{\lambda_t}$, then we obtain a rule which is called the *sequential probability ratio test*.

$$\tau_{(A,B)} = \inf\{n \geqq 0: \lambda_n(\omega) \notin (A,B)\},$$
$$d_{(A,B)} = \begin{cases} 1, & \lambda_{\tau_{(A,B)}} \geqq B, \\ 0, & \lambda_{\tau_{(A,B)}} \leqq A, \end{cases} \tag{4.94}$$

there is an optimal one. Moreover, there are examples that show that rules of the form (4.94) are in fact not optimal.

Here is one such example.

Suppose the densities (with respect to Lebesgue measure) $p_0(x)$ and $p_1(x)$ are given by the formulas

$$p_0(x) = \begin{cases} 1, & x \in [0, 1], \\ 0, & x \notin [0, 1], \end{cases}$$
$$p_1(x) = \begin{cases} 1, & x \in [a, a+1], \\ 0, & x \notin [a, a+1], \end{cases}$$

where $0 < a < 1$. Then for all $0 \leqq x \leqq a + 1$

$$\ln(p_1(x)/p_0(x)) = \begin{cases} \infty, & x \in [0, a], \\ 0, & x \in [a, 1], \\ -\infty, & x \in [1, 1+a]. \end{cases} \tag{4.95}$$

From (4.95) it is clear that for any choice of threshholds $A < 0 < B$ the error probabilities are

$$\mathbf{P}_0\{d_{(A,B)}(\omega) = 1\} = 0, \qquad \mathbf{P}_1\{d_{(A,B)}(\omega) = 0\} = 0,$$

where

$$\mathbf{E}_0\tau_{(A,B)} = \mathbf{E}_1\tau_{(A,B)} = 1/a.$$

Therefore if $\alpha > 0$ and $\beta > 0$, then for every choice of constants A and B, $A < 0 < B$, it cannot be that the error probabilities of first and second kind are exactly equal to $\alpha > 0$ and $\beta > 0$.

At the same time, rules distinct from $\delta_{(A,B)}$, for which the error probabilities are equal to the given values $\alpha > 0$ and $\beta > 0$, exist, and for them the mathematical expectations of the observation time (for each of the hypotheses H_1 and H_0) are less than a^{-1}.

Thus, for example, suppose the rule $\delta_h = (\tau_h, d_h)$ is such that

$$\tau_h(\omega) = \inf\{\pi \geqq 1: \xi_n(\omega) \notin (a+h, 1-h)\},$$
$$d_h(\omega) = \begin{cases} 1, & \xi_{\tau_n}(\omega) \in [1-h, a+1], \\ 0, & \xi_{\tau_n}(\omega) \in [0, a+h], \end{cases}$$

where $0 < h < 1 - a/2$. Then

$$\alpha(\delta_h) = \mathbf{P}_1\{d_h(\omega) = 1\} = h, \qquad \beta(\delta_h) = \mathbf{P}_0\{d_h(\omega) = 0\} = h$$

and

$$\mathbf{E}_0\tau_h = \mathbf{E}_1\tau_h = 1/(a+2h) < 1/a = \mathbf{E}_0\tilde{\tau} = \mathbf{E}_1\tilde{\tau}.$$

6. In the case of a Wiener process for given α and β we can find threshholds $\tilde{A}$ and $\tilde{B}$ exactly determining the optimal rule, as well as the mathematical expectations $\mathbf{E}_0\tilde{\tau}$ and $\mathbf{E}_1\tilde{\tau}$ of the time $\tilde{\tau}$ needed for testing hypotheses with error probabilities not exceeding α and β. For discrete time the problem of finding $\tilde{A}$, $\tilde{B}$, $\mathbf{E}_0\tilde{\tau}$ and $\mathbf{E}_1\tilde{\tau}$ in the general case is very difficult. We can, however, give bounds for these quantities which as a rule turn out to be completely satisfactory for applications.

Suppose $\delta_{(A,B)} = (\tau_{(A,B)}, d_{(A,B)})$ is a decision rule such that

$$\tau_{(A,B)} = \inf\{n \geqq 1 : \lambda_n \notin (A, B)\},$$
$$d_{(A,B)} = \begin{cases} 1, & \lambda_{\tau_{(A,B)}} \geqq B, \\ 0, & \lambda_{\tau_{(A,B)}} \leqq A, \end{cases} \tag{4.96}$$

where

$$\lambda_n = \ln \varphi_n = \sum_{k=1}^{n} \ln \frac{p_1(\xi_k)}{p_0(\xi_k)}.$$

As follows from Theorem 4, the optimal rule $\tilde{\delta}$ is a rule of the type (4.96). We let

$$\alpha(A,B) = \alpha(\delta_{(A,B)}) = \mathbf{P}_1\{d_{(A,B)}(\omega) = 0\}$$

and

$$\beta(A,B) = \beta(\delta_{(A,B)}) = \mathbf{P}_0\{d_{(A,B)}(\omega) = 1\}.$$

THEOREM 5. *If for given constants A and B*

$$\mathbf{P}_i\{\tau_{(A,B)} < \infty\} = 1, \qquad i = 0, 1, \tag{4.97}$$

and $\alpha(A,B) < 1$ and $\beta(A,B) < 1$, then

$$\ln \frac{\alpha(A,B)}{1-\beta(A,B)} \leqq A, \qquad B \leqq \ln \frac{1-\alpha(A,B)}{\beta(A,B)}. \tag{4.98}$$

PROOF. For given A and B the probability

$$\begin{aligned}\alpha(A,B) &= \mathbf{P}_1\{\lambda_{\tau_{(A,B)}} \leqq A\} = \int_{\{\lambda_{\tau_{(A,B)}} \leqq A\}} d\mathbf{P}_1(\omega) \\ &= \int_{\{\lambda_{\tau_{(A,B)}} \leqq A\}} \frac{p_1(\xi_1)\cdots p_1(\xi_{\tau_{(A,B)}})}{p_0(\xi_1)\cdots p_0(\xi_{\tau_{(A,B)}})}\, d\mathbf{P}_0(\omega) \\ &= \int_{\{\lambda_{\tau_{(A,B)}} \leqq A\}} e^{\lambda_{\tau_{(A,B)}}} d\mathbf{P}_0(\omega) \\ &\leqq e^A \mathbf{P}_0\{\lambda_{\tau_{(A,B)}} \leqq A\} = e^A[1-\beta(A,B)],\end{aligned}$$

whence follows the first inequality of (4.98). The validity of the second inequality in (4.98) can also be established in an analogous fashion.

7. In the theorem below we give conditions which, in particular, guarantee satisfaction of the requirements (4.97) involved in the statement of Theorem 5.

THEOREM 6. *Suppose* $-\infty < A \leqq 0 \leqq B < \infty$ *and*

$$\mathbf{P}_i\{|\ln(p_1(\xi_1)/p_0(\xi_0))| > 0\} > 0, \qquad i = 0, 1.$$

Then $\mathbf{P}_i\{\tau_{(A,B)} > \infty\} = 1$ *and there exists* $t_0 > 0$ *such that for all* $t \leqq t_0$

$$\mathbf{E}_i \exp[t_{\tau_{(A,B)}}] < \infty, \qquad i = 0, 1. \tag{4.99}$$

PROOF. Let

$$z_k = \ln(p_1(\xi_k)/p_0(\xi_k)), \qquad s_k = z_1 + \cdots + z_k$$

and $C = B - A$. We first assume that $\mathbf{P}_i\{|z_k| \leqq C\} = p_i < 1$. Then

$$\begin{aligned}\{\omega : \tau_{(A,B)} \geqq n\} &= \{\omega : A < s_k < B, 1 \leqq k \leqq n-1\} \\ &\subseteq \{\omega : |z_k| \leqq C, 1 \leqq k \leqq n-1\}\end{aligned}$$

and consequently

$$\mathbf{P}_i\{_{(A,B)} \geqq \tau\, n\} \leqq p_i^{n-1}, \tag{4.100}$$

whence

$$\begin{aligned}\mathbf{E}_i e^{t\tau_{(A,B)}} &= \sum_{k=1}^{\infty} e^{tk}\mathbf{P}_i\{\tau_{(A,B)} = k\} \\ &\leqq \sum_{k=1}^{\infty} e^{tk}\mathbf{P}_i\{\tau_{(A,B)} \geqq k\} \leqq e^t \sum_{k=0}^{\infty}(e^t p_i)^k < \infty,\end{aligned}$$

if $e^t p_i < 1$.

Now suppose $\mathbf{P}_i\{|z_k| \leqq C\} = 1$. Then there is a finite $m \geqq 1$ such that

$$\mathbf{P}_i\{|z_1 + \cdots + z_k| \leqq C\} = p_i < 1, \qquad p_i > 0,$$

whence it follows that

$$\mathbf{P}_i\{\tau_{(A,B)} \geqq mk\} \leqq p_i^{m-1}$$

and

$$\mathbf{P}_i\{\tau^{(A,B)} \geqq n\} \leqq p_i^{[n/m]-1} \leqq p_i^{-2}(p_i^{1/m})^n. \tag{4.101}$$

Consequently

$$\mathbf{E}_i e^{t\tau_{(A,B)}} \leqq p_i^{-2} \sum_{k=1}^{\infty} (e^t p_i^{1/m})^k < \infty,$$

provided $e^t p_i^{1/m} < 1$.

From the inequality (4.99) it follows, finally, that not only $\mathbf{P}_i\{\tau_{(A,B)} < \infty\} = 1$ but also that the times $\mathbf{E}_i\tau^n_{(A,B)} < \infty$ for all $n \geqq 1$.

8. The derivation of lower bounds for the mean number of observations required is essentially based on one general result, known as Wald's identity (cf. Lemma 8).

Suppose $(\Omega,\mathscr{F},\mathbf{P})$ is a probability space and $\xi,\xi_1,\xi_2,\cdots$ is a sequence of independent identically distributed random variables. Let $\mathscr{F}_n = \sigma\{\omega:\xi_1,\cdots,\xi_n\}$ and $s_n = \xi_1 + \cdots + \xi_n$, and let $\tau = \tau(\omega)$ denote a Markov time (relative to the system $F = \{\mathscr{F}_n, n \geqq 1\}$) taking on the values $1,2,\cdots$.

LEMMA 10 (WALD'S IDENTITY). *If* $\mathbf{E}|\xi| < \infty$ *and* $\mathbf{E}\tau < \infty$, *then*

$$\mathbf{E}s_\tau = \mathbf{E}\xi \cdot \mathbf{E}\tau. \tag{4.102}$$

If moreover $\mathbf{E}\xi^2 < \infty$, *then*

$$\mathbf{E}[s_\tau - \tau\mathbf{E}\xi]^2 = \mathrm{Var}\xi \cdot \mathbf{E}\tau, \tag{4.103}$$

where $\mathrm{Var}\xi = \mathbf{E}\xi^2 - (\mathbf{E}\xi)^2$.

PROOF. Let $\tau_N = \min(\tau,N)$, where $N < \infty$. Let $\eta_n = s_n - n\mathbf{E}\xi$. It is easy to see that the elements $(\eta_n,\mathscr{F}_n,\mathbf{P})$, $n \geqq 1$, form a martingale.

It is obvious that

$$\mathbf{E}|\eta_{\tau_N}| < \infty, \ \lim_{n\to\infty} \int_{(\tau_N>n)} |\eta_n| d\mathbf{P} = 0. \tag{4.104}$$

Therefore we can apply Theorem I.6, by which $\mathbf{E}\eta_{\tau_N} = \mathbf{E}\eta_1 = 0$ and consequently

$$\mathbf{E}s_{\tau_N} = \mathbf{E}\xi \cdot \mathbf{E}_{\tau_N}. \tag{4.105}$$

Thus Wald's identity has been established for Markov times bounded with probability one. We proceed to the general case.

From (4.102) we have

$$\mathbf{E}\{|\xi_1| + \cdots + |\xi_{\tau_N}|\} = \mathbf{E}|\xi| \cdot \mathbf{E}\tau_N \leqq \mathbf{E}|\xi| \cdot \mathbf{E}\tau < \infty. \tag{4.106}$$

Since $\tau_N \uparrow \tau$, $\sum_{i=1}^{\tau_N}|\xi_i| \uparrow \sum_{i=1}^{\tau}|\xi_i|$ as $N \to \infty$ with probability one, then from (4.106) we obtain that

$$\mathbf{E}\{|\xi_1| + \cdots + |\xi_\tau|\} = \lim_{N\to\infty} \mathbf{E}\{|\xi_1| + \cdots + |\xi_{\tau_N}|\} \leqq \mathbf{E}|\xi| \cdot \mathbf{E}\tau < \infty.$$

Consequently

$$\begin{aligned} \mathbf{E}|\eta_t| &\leqq \mathbf{E}|s_\tau| + \mathbf{E}\tau \cdot \mathbf{E}|\xi| \\ &\leqq \mathbf{E}\{|\xi_1| + \cdots + |\xi_\tau|\} + \mathbf{E}\tau \cdot \mathbf{E}|\xi| < \infty. \end{aligned} \tag{4.107}$$

Now we shall show that

$$\lim_{n\to\infty} \int_{(\tau>n)} \eta|_n| \, d\mathbf{P} = 0. \tag{4.108}$$

We have

$$|\eta_n| \leqq |\xi_1 - \mathbf{E}\xi| + \cdots + |\xi_n - \mathbf{E}\xi| \leqq |\xi_1| + \cdots + |\xi_n| + n\mathbf{E}|\xi|.$$

On the set$\{\omega : \tau > n\}$

$$|\eta_n| \leqq |\xi_1| + \cdots + |\xi_\tau| + \tau\mathbf{E}|\xi|.$$

Therefore, since $\mathbf{E}(|\xi_1| + \cdots + |\xi_\tau|) < \infty$, $\mathbf{E}\tau < \infty$, we have

$$\int_{(\tau>n)} |\eta_n| d\mathbf{P} \leqq \int_{(\tau>n)} \{|\xi_1| + \cdots + |\xi_\tau|\} d\mathbf{P}$$

$$+ \mathbf{E}|\xi| \int_{(\tau>n)} \tau(\omega) d\mathbf{P}(\omega) \to 0 \qquad \text{as } n \to \infty,$$

which proves (4.108).

Theorem I.6 is applicable to the martingale $(\eta_n, \mathscr{F}_n, \mathbf{P})$ by conditions (4.107) and (4.108). Therefore $\mathbf{E}\eta_\tau = 0$ or, equivalently, $\mathbf{E}(s_\tau - \tau\mathbf{E}\xi) = 0$. But since $\mathbf{E}|s_\tau| < \infty$, $\mathbf{E}\tau < \infty$ and $\mathbf{E}|\xi| < \infty$, we obtain $\mathbf{E}s_\tau = \mathbf{E}\tau \cdot \mathbf{E}\xi$, which proves the first assertion of the lemma.

Noting that the process $(\eta_n^2 - n\text{Var}\xi, \mathscr{F}_n, \mathbf{P})$, $n \geqq 1$, is a martingale, we can prove the validity of (4.103) in a similar fashion.

The reasoning developed above, based on applying Theorem I.6, allows us also to obtain relationships (analogous to (4.102) and (4.103)) including moments of higher orders. For the case of bounded Markov times $\tau = \tau(\omega)$ ($\mathbf{P}(\tau \leqq N) = 1$, $N < \infty$), they can be most easily obtained from the fundamental indentity of sequential analysis:

$$\mathbf{E}\{e^{\lambda s_\tau}[\varphi(\lambda)]^{-\tau}\} = 1, \tag{4.109}$$

where the complex number λ is such that $\varphi(\lambda) = \mathbf{E}e^{\lambda\xi}$ exists and does not vanish.

Formula (4.109) follows directly from Theorem I.6, if we note that the elements $(e^{\lambda s_n}[\varphi(\lambda)]^{-n}, \mathscr{F}_n, \mathbf{P})$, $n \geqq 1$, form a martingale and

$$\mathbf{E}e^{\lambda s_1}[\varphi(\lambda)]^{-1} = 1. \tag{4.110}$$

By a limiting passage from "truncated" times $\tau_N = \min(\tau, N)$ we can sometimes also establish the identity (4.109) for Markov times τ belonging to the class $\mathfrak{M}$ (see the Remark to Lemma 9).

9. Now we present estimates for the mean number of needed observations in problems of discriminating among N alternative hypotheses where, generally speaking, we assume $N \geqq 2$. Since we considered the case of only two hypotheses above, we need to introduce some additional notation called for by this assumption.

Suppose that on the measurable space $(\Omega, \mathscr{F})$ we are given measures $\mathbf{P}_\theta$, $\theta = 0$,

$1,\cdots, N-1$, and a sequence of independent identically distributed (with respect to each measure $\mathbf{P}_\theta$) random variables $\xi_1,\xi_2,\cdots$. We can assume without loss of generality that the probability distributions $\mathbf{P}_\theta(x) = \mathbf{P}_\theta\{\omega:\xi_1(\omega) \leqq x\}$ have densities $p_\theta(x)$ relative to some (σ-finite) measure μ.

We set $\mathscr{F}_n = \sigma\{\omega:\xi_1,\cdots, \xi_n\}$, $n \geqq 1$, and let $\tau = \tau(\omega)$ be a Markov time (relative to the system $\{\mathscr{F}_n\}$, $n \geqq 1$), taking on the values $1,2,\cdots,\infty$, such that $\mathbf{P}_\theta\{\tau < \infty\} = 1$ for all $\theta = 0,1,\cdots,N-1$. Also let $d = d(\omega)$be an $\mathscr{F}_\tau$-measurable function taking on N values $d_0,\cdots,d_{N-1}$. The value $d(\omega) = d_i$ is interpreted as accepting the hypothesis $H_i:\theta = i$.

Let

$$\alpha_{ij} = \mathbf{P}_i\{d(\omega) = d_j\}, \qquad 0 \leqq i,j \leqq N-1,$$

be the probability of accepting the hypothesis H_j when $\theta = i$ and the decision rule $\delta = (\tau,d)$ is used.

THEOREM 7. *Suppose i, $0 \leqq i \leqq N-1$, is fixed and $\alpha_{ik} = 0$ whenever the probability $\alpha_{jk} = 0$ for some $j \neq i$. Also suppose $\mu\{x:p_i(x) \neq p_j(x)\} > 0$, $j \neq i$. Then*

$$\mathbf{E}_i\tau \geqq \max_{j\neq i} \frac{\sum_{k=0}^{N-1}\alpha_{ik} \ln\dfrac{\alpha_{ik}}{\alpha_{jk}}}{\mathbf{E}_i\ln \dfrac{p_i(\xi_1)}{p_j(\xi_1)}}, \tag{4.111}$$

where expressions of the form $0\cdot \ln(0/c)$ are considered equal to zero for all $c \geqq 0$.

The proof is carried out according to the same scheme as the proof of the inequalities (4.78). If $\mathbf{E}_i\tau = \infty$, inequality (4.111) is obvious. Thus below we assume $\mathbf{E}_i\tau < \infty$.

Along with i, we also fix some $j \neq i$. Then, by Jensen's inequality and the assumption $\mu\{x: p_i(x) \neq p_j(x)\} > 0$,

$$0 < \mathbf{E}_i\ln\frac{p_i(\xi_1)}{p_j(\xi_1)} = \int_{\{x:p_i(x)>0\}} \ln\frac{p_i(x)}{p_j(x)}p_i(x)d\mu(x), \tag{4.112}$$

(see [44], Russian p. 63). If $\mathbf{E}_i \ln(p_i(\xi_1)/p_j(\xi_1)) = \infty$, then the inequality (4.111) is trivial.

Suppose $\mathbf{E}_i \ln(p_i(\xi_1)/p_j(\xi_1)) < \infty$. According to Lemma 10,

$$\mathbf{E}_i\ln\frac{p_i(\xi_1) \cdots p_i(\xi_\tau)}{p_j(\xi_1) \cdots p_j(\xi_\tau)} = \mathbf{E}_i\tau \cdot \mathbf{E}_i\ln\frac{p_i(\xi_1)}{p_j(\xi_1)}.$$

Letting

$$\eta(\omega) = \frac{p_j(\xi_1(\omega))\cdots p_j(\xi_\tau(\omega))}{p_i(\xi_1(\omega))\cdots p_i(\xi_\tau(\omega))} \quad\text{and}\quad D_i = \{0 \leqq k \leqq N-1:\alpha_{ik} \neq 0\},$$

from Jensen's inequality for $\ln \eta(\omega)$ we obtain

$$\begin{aligned}
\mathbf{E}_i \ln \frac{p_i(\xi_1)\cdots p_i(\xi_\tau)}{p_j(\xi_1)\cdots p_j(\xi_\tau)} &= -\mathbf{E}_i \ln \eta(\omega) \\
&= -\sum_{k=0}^{N-1} \int_{\{\omega: d(\omega)=d_k\}} \ln \eta(\omega) d\mathbf{P}_i(\omega) \\
&= -\sum_{k\in D_i} \int_{\{\omega: d(\omega)=d_k\}} \ln \eta(\omega) d\mathbf{P}_i(\omega) \\
&= -\sum_{k\in D_i} \mathbf{P}_i\{d(\omega)=d_k\} \int_\Omega \ln \eta(\omega) d\mathbf{P}_i(\omega | d(\omega)=d_k) \qquad (4.113)\\
&\geqq -\sum_{k\in D_i} \mathbf{P}_i\{d(\omega)=d_k\} \ln \int_\Omega \eta(\omega) d\mathbf{P}_i(\omega | d(\omega)=d_k) \\
&= -\sum_{k\in D_i} \mathbf{P}_i\{d(\omega)=d_k\} \\
&\quad \times \ln\left\{\sum_{n=1}^{\infty} \int_{A_{kn}} \frac{p_j(\xi_1)\cdots p_j(\xi_n)}{p_i(\xi_1)\cdots p_i(\xi_n)} \cdot \frac{d\mathbf{P}_i(\omega)}{\mathbf{P}_i\{d(\omega)=d_k\}}\right\},
\end{aligned}$$

where

$$A_{kn} = \{\tau = n\} \cap \{p_i(\xi_1)\cdots p_i(\xi_n) > 0\} \cap \{d(\omega) = d_k\}.$$

But

$$\begin{aligned}
&\int_{A_{kn}} \frac{p_j(\xi_1)\cdots p_j(\xi_n)}{p_i(\xi_1)\cdots p_i(\xi_n)} \frac{d\mathbf{P}_i(\omega)}{\mathbf{P}_i(d(\omega)=d_k)} \qquad (4.114)\\
&\quad = \frac{1}{\alpha_{ik}} \int_{A_{kn}} d\mathbf{P}_j(\omega) \leqq \frac{1}{\alpha_{ik}} \int_{\{\tau=n\} \cup \{d=d_k\}} d\mathbf{P}_j(\omega).
\end{aligned}$$

Therefore it follows from (4.113) and (4.114) that

$$\begin{aligned}
\mathbf{E}_i \ln \frac{p_i(\xi_1)\cdots p_i(\xi_\tau)}{p_j(\xi_1)\cdots p_j(\xi_\tau)} &\geqq -\sum_{k\in D_i} \alpha_{ik} \ln\left\{\frac{1}{\alpha_{ik}} \int_{\{d=d_k\}} d\mathbf{P}_j(\omega)\right\} \\
&= -\sum_{k\in D_i} \alpha_{ik} \ln \frac{\alpha_{jk}}{\alpha_{ik}} = \sum_{k\in D_i} \alpha_{ik} \ln \frac{\alpha_{ik}}{\alpha_{jk}} = \sum_{k=0}^{N-1} \alpha_{ik} \ln \frac{\alpha_{ik}}{\alpha_{jk}}. \qquad (4.115)
\end{aligned}$$

Comparing this inequality with (4.112), we obtain the bound

$$\mathbf{E}_i \tau \geqq \frac{\sum_{k=0}^{N-1} \alpha_{ik} \ln \dfrac{\alpha_{ik}}{\alpha_{jk}}}{\mathbf{E}_i \ln \dfrac{p_i(\xi_1)}{p_j(\xi_1)}}, \qquad (4.116)$$

from which (4.111) follows immediately.

COROLLARY. *Let $N = 2$. Letting $\alpha = \alpha_{10}$, $\beta = \alpha_{01}$, from* (4.111) *we find*

$$\mathbf{E}_0 \tau \geqq \frac{\omega(\beta,\alpha)}{\mathbf{E}_0 \ln(p_0(\xi_1)/p_1(\xi_1))}, \qquad \mathbf{E}_1 \tau \geqq \frac{\omega(\alpha,\beta)}{\mathbf{E}_1 \ln(p_1)\xi_1)/p_0(\xi_1))},$$

where

$$\omega(x,y) = (1 - x)\ln\frac{1 - x}{y} + x \ln\frac{x}{1 - y}.$$

COROLLARY 2. *Let* $N \geqq 2$, $\alpha_{ii} = \alpha$ *for all* $i = 0,1,\cdots, N-1$ *and* $\alpha_{ij} - (1-\alpha)/(N-1)$, $i \neq j$. *Then*

$$\mathbf{E}_i\tau \geqq \frac{\dfrac{\alpha N-1}{N-1}\ln\dfrac{\alpha(N-1)}{1-\alpha}}{\min\limits_{j\neq i} \mathbf{E}_i\ln\dfrac{p_i}{p_j}}. \tag{4.117}$$

§ 3. Disruption problem. Discrete time

1. In the problem of testing two simple hypotheses considered above, the one-dimensional probability distribution of the random variables $\xi_1, \xi_2, \cdots$ remained unchanged (although unknown) during the entire process of observation.

In detection theory and statistical control we often encounter problems in which the probabilistic characteristics of the variables being observed may change at a random moment of time $\theta = \theta(\omega)$ (the time of appearance of a "disruption"). Below we present several such problems and propose a method for solving them based on the general theory of optimal stopping rules set forth in the preceding two chapters.

In order to give a more precise formulation of the problems, we need some notation.

Suppose $(\Omega,\mathscr{F})$ is a measurable space and $\{\mathbf{P}_\pi, 0 \leqq \pi \leqq 1\}$ is a family of probability measures on it. We assume that on $\{\Omega,\mathscr{F}\}$ we are given: a random variable $\theta = \theta(\omega)$ with values in $\boldsymbol{N} = \{0,1,\cdots\}$, and a sequence of random variables $\xi_1,\xi_2,\cdots$. Suppose that for each $0 \leqq \pi \leqq 1$

$$\mathbf{P}_\pi\{\theta(\omega) = 0\} = \pi, \qquad \mathbf{P}_\pi\{\theta(\omega) = n \mid \theta(\omega) > 0\} = (1-p)^{n-1}p, \tag{4.118}$$

where the constant $0 < p < 1$ is considered unknown and not dependent on π.

Further suppose that for every $n \geqq 1$ the probabilities

$$P_\pi(x_1, \cdots, x_n) = \mathbf{P}_\pi\{\xi_1 \leqq x_1, \cdots, \xi_n \leqq x_n\}$$

are such that

$$\begin{aligned} P_\pi(x_1, \cdots, x_n) &= \pi P_1(x_1, \cdots, x_n) \\ &+(1-\pi)\sum_{i=0}^{n-1} p(1-p)^i P_0(x_1, \cdots, x_i)P_1(x_{i+1}, \cdots, x_n) \\ &+(1-\pi)(1-p)^n P_0(x_1, \cdots, x_n). \end{aligned} \tag{4.119}$$

We also assume that the random variables $\xi_1,\xi_2,\cdots$ are mutually independent with respect to each of the measures $\mathbf{P}_0$ and $\mathbf{P}_1$, i.e. for all $n \geqq 1$

$$P_i(x_1, \cdots, x_n) = P_i(x_1)\cdots P_i(x_n), \qquad i = 0,1. \tag{4.120}$$

Without loss of generality we can assume that the distributions $P_i(x)$ have density $p_i(x)$, $i = 0,1$ (with respect to some σ-finite measure μ).

The intuitive meaning of the conditions (4.118)—(4.120) is the following. If $\theta(\omega) = 0$, then we are observing a sequence of independent identically distributed random variables $\xi_1, \xi_2, \cdots$ with probability density $p_1(x)$. But under the condition $\theta(\omega) = i$ the random variables $\xi_1, \cdots, \xi_{i-1}, \xi_i, \cdots$ are mutually independent and $\xi_1, \cdots, \xi_{i-1}$ are identically distributed with probability density $p_0(x)$, while $\xi_i, \xi_{i+1}, \cdots$ are also identically distributed, but with probability density $p_1(x)$, The formulation considered below of the problem of quickest detection of the time of appearance of a "disruption" is also meaningful for distributions more general than (4.118). The geometric nature of the probability distribution of the time of appearance of a disruption is used by us only for the sake of simplicity of presentation.

2. Let $\xi_0(\omega) \equiv 0$ and $\mathscr{F}_n = \sigma\{(\omega,\pi): \pi, \xi_0(\omega), \cdots, \xi_n(\omega)\}$, $n \geqq 0$, and suppose $\mathfrak{M} = \{\tau\}$ is the class of stopping times $\tau = \tau(\omega,\pi)$ (relative to the system $F = \{\mathscr{F}_n\}, n \geqq 0$).

To each Markov time $\tau = \tau(\omega,\pi)$ corresponds a probability of "false alarm" $\mathbf{P}_\pi(\tau < \theta)$ and a mean delay time $\mathbf{E}_\pi(\tau - \theta | \tau \geqq \theta)$ of detecting the time of appearance of disruptions under the assumption that the "alarm" signal is correctly given, i.e. under the condition $\{\tau \geqq \theta\}$.

It is natural to wish to find a Markov time such that both the probability of a "false alarm" as well as the mean delay time are as small as possible. The contradictory nature of these requirements leads us to the following formulation.

Variational formulation. Suppose π is fixed, $0 \leqq \pi < 1$, and $\mathfrak{M}(\alpha;\pi) \subseteq \mathfrak{M}$ is the class of Markov times $\tau = \tau(\omega,\pi)$ for which

$$\mathbf{P}_\pi(\tau < \theta) \leqq \alpha, \tag{4.121}$$

where α is a given constant, $0 \leqq \alpha < 1$. The time $\bar{\tau} = \bar{\tau}(\omega,\pi) \in \mathfrak{M}(\alpha;\pi)$ is said to be optimal if

$$\mathbf{E}_\pi(\bar{\tau} - \theta \mid \bar{\tau} \geqq \theta) \leqq \mathbf{E}_\pi(\tau - \theta \mid \tau \geqq \theta) \tag{4.122}$$

for all $\tau \in \mathfrak{M}(\alpha; \pi)$.

In Theorem 9 we give conditions under which an optimal rule exists.

The proof of this theorem is based essentially on knowing the best solution in the following (Bayesian) formulation of the "disruption" problem.

Bayesian formulation. Let

$$\rho(\pi, \tau) = \mathbf{P}_\pi(\tau < \theta) + c\mathbf{E}_\pi(\tau - \theta \mid \tau \geqq \theta)\mathbf{P}_\pi(\tau \geqq \theta),$$

where the constant $c > 0$, and $\rho(\pi) = \inf_{\tau \in \mathfrak{M}} \rho(\pi, \tau)$.

The Markov time $\tau^* = \tau^*(\omega,\pi)$ is said to be π-*Bayesian* if for given π

$$\rho(\pi, \tau^*) = \rho(\pi). \tag{4.123}$$

If for some $\tau^* = \tau^*(\omega,\pi)$ the equality (4.123) is satisfied for all π, $0 \leqq \pi \leqq 1$, then we call this time *Bayesian.*

3. THEOREM 8. *Suppose $c > 0$ and $p > 0$, and let*

$$\pi_n(\omega, \pi) = \mathbf{P}_\pi\{\theta(\omega) \leqq n \mid \mathscr{F}_n\}, \qquad n \geqq 1,$$

be the posterior probability of the presence of a disruption at the moment of time n; set $\pi_0(\omega,\pi) = \pi$. Then the time

$$\tau^*(\omega,\pi) = \inf\{n \geqq 0: \pi_n(\omega,\pi) \geqq A^*\},$$

where A^* is some constant, is Bayesian.

PROOF. As in § 1, we first show that finding a Bayesian time can be reduced to solving a special problem on optimal stopping for some Markov process.

By Bayes' formula we have $\mathbf{P}_\pi$-a.s. for all $n \geqq 0$

$$\pi_{n+1}(\omega, \pi) = \frac{\pi_n p_1(\xi_{n+1}) + (1 - \pi_n)p\cdot p_1(\xi_{n+1})}{\pi_n p_1(\xi_{n+1}) + (1 - \pi_n)p\cdot p_1(\xi_{n+1}) + (1 - \pi_n)(1 - p)p_0(\xi_{n+1})}, \tag{4.124}$$

where $\pi_n = \pi_n(\omega, \pi)$ and $\xi_{n+1} = \xi_{n+1}(\omega)$. Hence by Lemma 2 it follows that the elements $\Pi_\pi = \{\pi_n(\omega, \pi), \mathscr{F}_n, \mathbf{P}_\pi\}$ form a Markov random function.

For each $0 \leqq \pi \leqq 1$ and $\tau \in \mathfrak{M}$

$$\begin{aligned} \rho(\pi, \tau) &= \mathbf{P}_\pi(\tau < \theta) + c\mathbf{E}_\pi(\tau - \theta \mid \tau \geqq \theta)\mathbf{P}_\pi(\tau \geqq \theta) \\ &= \mathbf{E}_\pi(1 - \pi_\tau) + c\mathbf{E}_\pi \max(\tau - \theta, 0), \end{aligned} \tag{4.125}$$

where $\mathbf{E}_\pi$ denotes averaging with respect to the measure $\mathbf{P}_\pi$. We transform the mathematical expectation $\mathbf{E}_\pi \max(\tau - \theta, 0)$ to a form more suited to our purpose.

For each $n \geqq 0$ we have

$$\begin{aligned} &\mathbf{E}_\pi[\max(n - \theta, 0) \mid \mathscr{F}_n] \\ &= \sum_{k=0}^{n} (n - k)\mathbf{P}_\pi(\theta = k \mid \mathscr{F}_n) = \sum_{k=0}^{n-1} [\mathbf{P}_\pi(\theta \leqq k \mid \mathscr{F}_n)] \\ &= \sum_{k=0}^{n-1} [\mathbf{P}_\pi(\theta \leqq k \mid \mathscr{F}_n) - \mathbf{P}_\pi(\theta \leqq k \mid \mathscr{F}_k)] \\ &\quad + \sum_{k=0}^{n-1} \mathbf{P}_n(\theta \leqq k \mid \mathscr{F}_k) \\ &= \sum_{k=0}^{n-1} [\mathbf{P}_\pi(\theta \leqq k \mid \mathscr{F}_n) - \mathbf{P}_\pi(\theta \leqq k \mid \mathscr{F}_k)] + \sum_{k=0}^{n-1} \pi_k. \end{aligned}$$

We let

$$\begin{aligned} \psi_n(\omega, \pi) &= \sum_{k=0}^{n} [\mathbf{P}_\pi(\theta \leqq k \mid \mathscr{F}_n) - \mathbf{P}_\pi(\theta \leqq k \mid \mathscr{F}_\pi)] \\ &= -\sum_{k=0}^{n} [\mathbf{P}_\pi(\theta > k \mid \mathscr{F}_n) - \mathbf{P}_\pi(\theta > k \mid \mathscr{F}_k)]. \end{aligned}$$

The sequence $(\psi_n(\omega,\pi),\mathscr{F}_n,\mathbf{P}_\pi)$, $n \geqq 0$, forms a martingale for every π with $0 \leqq \pi \leqq 1$. Indeed, it is obvious that $\mathbf{E}_\pi | \psi_n(\omega,\pi) | < \infty$ and

$$\psi_{n+1}(\omega,\pi) = \sum_{k=0}^{n} [\mathbf{P}_\pi(\theta \leqq k \mid \mathscr{F}_{n+1}) - \mathbf{P}_\pi(\theta \leqq k \mid \mathscr{F}_k)],$$

from which it follows that ($\mathbf{P}_\pi$-a.s.) $\mathbf{E}_\pi[\psi_{n+1}(\omega,\pi) \mid \mathscr{F}_n] = \psi_n(\omega,\pi)$.

Since

$$| \psi_n(\omega,\pi) | \leqq \sum_{k=0}^{\infty} \mathbf{P}_\pi(\theta > k \mid \mathscr{F}_n) + \sum_{k=0}^{\infty} \mathbf{P}_\pi(\theta > k \mid \mathscr{F}_k),$$

where (by the assumption $p > 0$)

$$\mathbf{E}_\pi \sum_{k=0}^{\infty} \mathbf{P}_\pi(\theta > k \mid \mathscr{F}_n) = \sum_{k=0}^{\infty} \mathbf{P}_\pi(\theta > k) = \mathbf{E}_\pi\theta < \infty$$

and

$$\mathbf{E}_\pi \sum_{k=0}^{\infty} \mathbf{P}_\pi(\theta > k \mid \mathscr{F}_k) = \mathbf{E}_\pi\theta < \infty,$$

it follows that for $\tau(\omega,\pi) \in \mathfrak{M}$

$$\lim_{n\to\infty} \int_{(\tau>n)} | \psi_n(\omega,\pi) | \, d\mathbf{P}_\pi = 0. \tag{4.126}$$

It is also clear that

$$\mathbf{E}_\pi | \psi_\tau(\omega,\pi) | < \infty, \qquad \tau \in \mathfrak{M}. \tag{4.127}$$

From conditions (4.126) and (4.127) it follows (see Theorem I. 6) that for every $\tau \in \mathfrak{M}$

$$\mathbf{E}_\pi\psi_\tau(\omega,\pi) = \mathbf{E}_\pi\psi_0(\omega,\pi) = 0, \qquad 0 \leqq \pi \leqq 1.$$

Therefore if $\tau \in \mathfrak{M}$, then

$$\begin{aligned}\rho(\pi,\tau) &= \mathbf{P}_\pi(\tau < \theta) + c\mathbf{E}_\pi(\tau - \theta \mid \tau \geqq \theta)\mathbf{P}_\pi(\tau \geqq \theta) \\ &= \mathbf{E}_\pi\left\{(1 - \pi_\tau) + c\sum_{k=0}^{\tau-1} \pi_k + c\psi_\tau(\omega,\pi)\right\} \\ &= \mathbf{E}_\pi\left\{(1 - \pi_\tau) + c\sum_{k=0}^{\tau-1} \pi_k\right\}\end{aligned}$$

and consequently

$$\rho(\pi) = \inf_{\tau\in\mathfrak{M}} \mathbf{E}_\pi\left\{(1 - \pi_\tau) + c\sum_{k=0}^{\tau-1} \pi_k\right\}. \tag{4.128}$$

The process $\Pi_\pi = (\pi_n(\omega,\pi),\mathscr{F}_n,\mathbf{P}_\pi)$, $n \geqq 0$, forms a sub-martingale ($\mathbf{E}_\pi[\pi_{n+1}|\mathscr{F}_n] \geqq \pi_n$, $\mathbf{P}_\pi$-a.s. $n \geqq 0$). Therefore (see Theorem I. 3) the limit $\lim_{n\to\infty} \pi_n$ exists with $\mathbf{P}_\pi$-probability one. Here it is obvious that $\lim_{n\to\infty} \pi_n \leqq 1$ and $\lim_{n\to\infty} \mathbf{E}\pi_n = 1$.

By Fatou's Lemma,

$$1 = \lim_{n\to\infty} \mathbf{E}\pi_n \leqq \mathbf{E} \lim_{n\to\infty} \pi_n,$$

and consequently $\lim_{n\to\infty} \pi_n = 1$ with $\mathbf{P}_\pi$-probability one for all π, $0 \leqq \pi \leqq 1$. Hence it follows that

$$\lim_{n\to\infty} \sum_{k=0}^{n} \pi_k = \infty \qquad (\mathbf{P}_\pi\text{-a.s., } 0 \leqq \pi \leqq 1). \tag{4.129}$$

In order to be able to apply the theory worked out above for optimal stopping rules, we associate a Markov process with the family of Markov random functions $\{\Pi_\pi, 0 \leqq \pi \leqq 1\}$.

Let $\Omega' = \Omega \times [0,1]$ be the space of points $\omega' = (\omega,\pi)$, let $\mathscr{F}' = \mathscr{F} \times \mathscr{B}([0, 1])$, and for all $n \geqq 1$ let

$$\pi_n'(\omega') = \pi_n(\omega,\pi)$$

and

$$\pi_0'(\omega') = \pi, \qquad \mathscr{F}_n' = \sigma\{\omega' : \pi_0'(\omega'), \xi_0'(\omega'),\dots,\xi_n'(\omega')\},$$

where $\xi_0'(\omega') \equiv 0$ and $\xi_n'(\omega') = \xi_n(\omega)$. We let $\mathbf{P}_\pi'$ denote the measure on the sets $A \in \mathscr{F}_\infty' = \sigma(\bigcup_{n\geqq 0} \mathscr{F}_n')$ defined by (4.15).

Then it can easily be established that the elements $\Pi' = (\pi_n',\mathscr{F}_n',\mathbf{P}_\pi')$, $n \in \boldsymbol{N}$, $\pi \in [0,1]$, form a Markov process. For simplicity we shall write π_n and $\mathscr{F}_n$ instead of π_n' and $\mathscr{F}_n'$.

Now we note that the class $\mathfrak{M}$ is also the class of Markov times relative to the system $F' = \{\mathscr{F}_n'\}$, $n \in \boldsymbol{N}$. Therefore, according to (4.128),

$$\begin{aligned} \rho(\pi) &= \inf_{\tau\in\mathfrak{M}} \mathbf{E}_\pi\Big\{(1 - \pi_\tau) + c\sum_{k=0}^{\tau-1} \pi_k\Big\} \\ &= \inf_{\tau\in\mathfrak{M}} \mathbf{E}_\pi'\Big\{(1 - \pi_\tau) + c\sum_{k=0}^{\tau-1} \pi_k\Big\}. \end{aligned} \tag{4.130}$$

Let $\mathfrak{M}^\pi$ be the class of Markov times $\tau \in \mathfrak{M}$ such that for each $n \in \boldsymbol{N}$

$$\{\omega' : \tau(\omega') = n\} \in \mathscr{F}_n^\pi,$$

where $\mathscr{F}_n^\pi = \sigma\{\omega' : \pi_0,\dots,\pi_n\}$. Then in view of the results of § 7.3 of Chapter II we have

$$\rho(\pi) = \inf_{\tau \in \mathfrak{M}} \mathbf{E}'_\pi \left\{ (1 - \pi_\tau) + c \sum_{k=0}^{\tau-1} \pi_k \right\}$$
$$= \inf_{\tau \in \mathfrak{M}^\pi} \mathbf{E}'_\pi \left\{ (1 - \pi_\tau) + c \sum_{k=0}^{\tau-1} \pi_k \right\}. \tag{4.131}$$

Thus the problem of finding a Bayesian time reduces to an optimal stopping problem for the Markov process Π', for whose solution we can apply the methods developed in § 8 of Chapter II.

Let $\mathfrak{N}^\pi$ denote the class of Markov times $\tau \in \mathfrak{M}^\pi$ for which

$$\mathbf{E}_\pi \sum_{k=0}^{\tau-1} \pi_k < \infty, \qquad 0 \leqq \pi \leqq 1.$$

Then (see § 3 of Chapter II)

$$\rho(\pi) \inf_{\tau \in \mathfrak{N}^\pi} \mathbf{E}'_\pi \left\{ (1 - \pi_\tau) + c \sum_{k=0}^{\tau-1} \pi_k \right\},$$

and consequently it now suffices to consider only Markov times τ from the class $\mathfrak{N}^\pi$.

Suppose $g(\pi) = 1 - \pi$ and

$$Q_c g(\pi) = \min\{g(\pi), c\pi + Tg(\pi)\},$$

where $Tg(\pi) = \mathbf{E}'_\pi g(\pi_1)$. Let

$$v(\pi) = \lim_{n \to \infty} Q_c^N g(\pi).$$

An insignificant modification of the proof of Lemma II.4 allows us to show that $-v(\pi)$ is the smallest $(1,c)$-excessive majorant (see § 8 of Chapter II) of the function $-g(\pi)$. Consequently, in particular,

$$- v(\pi) \geqq - c\pi - Tv(\pi), \quad 0 \leqq \pi \leqq 1,$$

and for all $n \in \boldsymbol{N}$

$$- v(\pi) \geqq - \mathbf{E}'_\pi \left\{ v(\pi_n) + c \sum_{k=0}^{n-1} \pi_k \right\}. \tag{4.132}$$

Starting with this inequality, we can easily establish (cf. the proof of inequality (1.43) and formula (2.122)) that for every bounded Markov time $\tau = \tau(\omega')$ $(\mathbf{P}'_\pi(\tau(\omega') \leqq N) = 1, 0 \leqq \pi \leqq 1, N < \infty)$

$$- v(\pi) \geqq - \mathbf{E}'_\pi \left\{ v(\pi_\tau) + c \sum_{k=0}^{\tau-1} \pi_k \right\}. \tag{4.133}$$

Hence it follows that if $\tau \in \mathfrak{N}^\pi$ and $\tau_N = \min(\tau, N)$, $N < \infty$, then

$$- v(\pi) \geqq - \int_{(\tau<N)} \left[v(\pi_\tau) + c \sum_{k=0}^{\tau-1} \pi_k \right] d\mathbf{P}'_\pi$$
$$- \int_{(\tau \geqq N)} \left[v(\pi_N) + c \sum_{k=0}^{N-1} \pi_k \right] d\mathbf{P}'_\pi.$$

But $| v(\pi) | \leqq 1$ and $\mathbf{E}'_\pi \sum_{k=0}^{\tau-1} \pi_k < \infty$ if $\tau \in \mathfrak{R}^\pi$; therefore as $N \to \infty$

$$\int_{(\tau \geqq N)} \left| v(\pi_N) + c \sum_{k=0}^{N-1} \pi_k \right| d\mathbf{P}'_\pi$$
$$\leqq \int_{(\tau \geqq N)} c \sum_{k=0}^{\tau-1} \pi_k d\mathbf{P}'_\pi + \mathbf{P}_\pi(\tau \geqq N) \to 0.$$

Consequently the function $-v(\pi)$ is the smallest regular $(1,c)$-excessive majorant of the function $-g(\pi)$.

From Theorem II.16 it follows that

$$\rho(\pi) = \lim_{n \to \infty} Q_c^N g(\pi) \tag{4.134}$$

and

$$\rho(\pi) = \min\{(1 - \pi), c\pi + T\rho(\pi)\}. \tag{4.135}$$

Simple verification shows that each of the functions $Q_c^N g(\pi)$ is convex. Therefore the risk $\rho(\pi)$ is also a convex function, continuous on the interval $(0,1)$

By (4.129) and the Corollary to Theorem II.16 for $c > 0$ the time

$$\tau^*(\omega') = \inf\{n \geqq 0 : \rho(\pi_n) = g(\pi_n)\}$$

is optimal in the class $\mathfrak{M}^\pi$, i.e.

$$\rho(\pi) = \inf_{\tau \in \mathfrak{M}^\pi} \mathbf{E}'_\pi \left\{ (1 - \pi_\tau) + c \sum_{k=0}^{\tau-1} \pi_k \right\}$$
$$= \mathbf{E}'_\pi \left\{ (1 - \pi_{\tau^*}) + c \sum_{k=0}^{\tau^*-1} \pi_k \right\}.$$

It is obvious that this time also belongs to the class $\mathfrak{R}^\pi$, i.e. $\mathbf{E}_\pi \sum_{\pi=0}^{\tau^*-1} \pi_k < \infty$ for all $0 \leqq \pi \leqq 1$.

Equation (4.135), the convexity and the continuity of each of the functions $1 - \pi$, $\rho(\pi)$ and $c\pi + T\rho(\pi)$ imply the existence of a number A^*, $0 \leqq A^* \leqq 1$, such that

$$\tau^*(\omega') = \inf\{n \geqq 0 : \rho(\pi_n) = g(\pi_n)\} = \inf\{n \geqq 0 : \pi_n \geqq A^*\}. \tag{4.136}$$

Consequently the time τ^* defined by formula (4.136) is a Bayesian time, which is what we wanted to prove.

4. We now proceed to finding an optimal time $\tilde{\tau} = \tilde{\tau}(\omega,\pi)$ in the variational formulation.

We fix some π, $0 \leqq \pi < 1$, and a constant α, $0 < \alpha < 1$. We let $\mathfrak{M}(\alpha;\pi)$ denote the collection of Markov times $\tau = \tau(\omega,\pi) \in \mathfrak{M}$ for which the probability of a false alarm $\mathbf{P}_\pi(\tau < \theta) \leqq \alpha$. First we note that for given π, only values $\alpha < 1 - \pi$ are of actual interest.

Indeed, if $\alpha \geqq 1 - \pi$, then, setting $\tilde{\tau}(\omega,\pi) \equiv 0$. we obtain

$$\mathbf{P}_\pi(\tilde{\tau} < \theta) = \mathbf{P}_\pi(\theta > 0) = 1 - \pi \leqq \alpha$$

and

$$\mathbf{E}_\pi \max(\tilde{\tau} - \theta, 0) = 0,$$

whence it follows that the time $\tilde{\tau}(\omega,\pi) \equiv 0$ is optimal in the class $\mathfrak{M}(\alpha;\pi)$ if $\alpha \geqq 1 - \pi$.

Thus we assume that $\alpha < 1 - \pi$. We let

$$\tau_A(\omega, \pi) = \inf\{n \geqq 0 \colon \pi_n(\omega, \pi) \geqq A\}$$

and

$$\alpha_A(\pi) = \mathbf{P}_\pi\{\tau_A(\omega, \pi) < \theta(\omega)\} = \mathbf{E}_\pi\{1 - \pi_{\tau_A(\omega,\pi)}\}.$$

It is clear that $\alpha_0(\pi) = 1 - \pi$, $\alpha_1(\pi) = 0$ and the function $\alpha_A(\pi)$ does not decrease as A increases, $0 \leqq A \leqq 1$. We further consider only the case in which $\alpha_A(\pi)$ is a continuous function in A.

Suppose $\alpha < 1 - \pi$ and $A(\alpha)$ is the smallest A for which $\alpha_A(\pi) = \alpha$. Consider the risk

$$\rho_c(\pi) = \inf_{\tau \in \mathfrak{M}} \{\mathbf{P}_\pi(\tau < \theta) + c\mathbf{E}_\pi(\tau - \theta \mid \tau \geqq \theta)\mathbf{P}_\pi(\tau \geqq \theta)\},$$

where the subscript c on $\rho_c(\pi)$ is introduced in order to emphasize the dependence of the risk on c. Let $A^*(c)$ be the value of the threshhold A^* (depending on c) which is involved in the definition of the Bayesian time (4.136). It can easily be shown that $A^*(c)$ is a continuous nonincreasing function of c, $A^*(0) = 1$ and $\lim_{c\to\infty} A^*(c) = 0$. We let $c^*(A^*)$ denote the minimal c for which $A^*(c)$ is equal to A^*.

Let $0 < \alpha < 1 - \pi$ and $c_\alpha = c^*(A(\alpha))$. Consider the risk

$$\rho_{c_\alpha}(\pi) = \inf_{\tau \in \mathfrak{M}} \{\mathbf{P}_\pi(\tau < \theta) + c_\alpha\mathbf{E}_\pi(\tau - \theta \mid \tau \geqq \theta)\mathbf{P}_\pi(\tau \geqq \theta)\}. \tag{4.137}$$

According to Theorem 8, the Bayesian time

$$\tau^*_{c_\alpha}(\omega, \pi) = \inf\{n \geqq 0 \colon \pi_n(\omega, \pi) \geqq A^*(c_\alpha)\}, \tag{4.138}$$

and, by the definition of c_α,

$$\mathbf{P}_\pi\{\tau^*_{c_\alpha}(\omega, \pi) < \theta\} = \alpha. \tag{4.139}$$

Let $\tau = \tau(\omega, \pi)$ be a Markov time belonging to the class $\mathfrak{M}(\alpha; \pi)$. Then

$$\begin{aligned} &\mathbf{P}_\pi(\tau < \theta) + c_\alpha \mathbf{E}_\pi(\tau - \theta \mid \tau \geqq \theta)\mathbf{P}_\pi(\tau \geqq \theta) \\ &\quad \geqq \mathbf{P}_\pi(\tau^*_{c_\alpha} < \theta) + c_\alpha \mathbf{E}_\pi(\tau^*_{c_\alpha} - \theta \mid \tau^*_{c_\alpha} \geqq \theta)\mathbf{P}_\pi(\tau^*_{c_\alpha} \geqq \theta) \\ &\quad = \alpha + c_\alpha(1 - \alpha)\mathbf{E}_\pi(\tau^*_{c_\alpha} - \theta \mid \tau^*_{c_\alpha} \geqq \theta). \end{aligned}$$

But $\mathbf{P}_\pi(\tau < \theta) \leqq \alpha$, and therefore

$$\begin{aligned} &c_\alpha(1 - \alpha)\mathbf{E}_\pi(\tau - \theta \mid \tau \geqq \theta) \\ &\quad \geqq c_\alpha \mathbf{P}_n\{\tau^*_{c_\alpha} \geqq \theta\}\mathbf{E}_\pi(\tau^*_{c_\alpha} - \theta \mid \tau^*_{c_\alpha} \geqq \theta). \end{aligned} \tag{4.140}$$

If $0 < \alpha < 1 - \pi$, then $c_\alpha \neq 0$. Indeed, if $c_\alpha = 0$, then $\mathbf{P}_\pi\{\tau_0^* < \theta\} = \mathbf{E}_\pi\{1 - \pi_{\tau_0^*}\} = 0$.

Thus $c_\alpha(1 - \alpha) > 0$, and from (4.140) we obtain that for all $\tau \in \mathfrak{M}(\alpha;\pi)$, $0 < \alpha < 1 - \pi$,

$$\mathbf{E}_\pi(\tau - \theta \mid \tau \geqq \theta) \geqq \mathbf{E}_\pi(\tau^*_{c_\alpha} - \theta \mid \tau^*_{c_\alpha} \geqq \theta). \tag{4.141}$$

Finding the exact value of the variable $\tilde{A} = A^*(c_\alpha)$ for each $0 < \alpha < 1$ is a very difficult task. Therefore the following bound for $\tilde{A}$ may turn out to be useful:

$$\tilde{A} \leqq 1 - \alpha. \tag{4.142}$$

To prove (4.142) we need note only that for each $0 \leqq A \leqq 1$ and $\tau_A(\omega,\pi) = \min\{n \geqq 0: \pi_n(\omega,\pi) \geqq A\}$

$$\mathbf{E}_\pi\{1 - \pi_{\tau_A}\} \leqq 1 - A,$$

whence $A \leqq 1 - \mathbf{P}_\pi(\tau_A < \theta)$, which proves (4.142).

Thus we have proved

THEOREM 9. *Let $0 < \alpha < 1$, $0 \leqq \pi < 1$ and $p > 0$, and suppose $\mathfrak{M}(\alpha;\pi)$ is the collection of Markov times τ for which $\mathbf{P}_\pi(\tau < \theta) \leqq \alpha$. If the function $\alpha_A(\pi)$ is continuous in A for all π, then the Markov time*

$$\tilde{\tau} = \tau^*_{c_\alpha} = \inf\{n \geqq 0: \pi_n \geqq \tilde{A}\}, \tag{4.143}$$

where $\tilde{A} = A^*(c_\alpha)$ is optimal in the sense that for each $\tau \in \mathfrak{M}(\alpha;\pi)$

$$\mathbf{E}_\pi(\tilde{\tau} - \theta \mid \tilde{\tau} \geqq \theta) \leqq \mathbf{E}_\pi(\tau - \theta \mid \tau \geqq \theta).$$

The threshhold $\tilde{A} = A^*(c_\alpha)$ satisfies the inequality (4.152).

REMARK. The function $\alpha_A(\pi)$ is continuous in A if for each n the probability distribution function of the random variables $\pi_n(\omega,\pi)$ is continuous. In turn this condition is satisfied if, say, the densities $p_0(x)$ and $p_1(x)$ (with respect to Lebesgue measure) are Gaussian.

§ 4. Disruption problem for a Wiener process

1. Suppose that on the measurable space $(\Omega, \mathscr{F})$ we are given:
1) a family of probability measures $\{\mathbf{P}_\pi, 0 \leqq \pi \leqq 1\}$;
2) a random variable $\theta = \theta(\omega)$ with values in $[0, \infty)$ such that

$$\mathbf{P}_\pi\{\theta(\omega) = 0\} = \pi, \quad \mathbf{P}_\pi\{\theta(\omega) \geqq t \,|\, \theta(\omega) > 0\} = e^{-\lambda t}, \qquad t > 0, \tag{4.144}$$

where the constant λ, $0 < \lambda < \infty$, is known and does not depend on π;
3) a standard Wiener process $\{w_t\}$, $t \geqq 0$, which is independent of $\theta = \theta(\omega)$ for each of the measures $\mathbf{P}_\pi$ and is such that $w_0(\omega) \equiv 0$, $\mathbf{E}_\pi \Delta w_t = 0$ and $\mathbf{E}_\pi \Delta w^2_t = \Delta t$.

Suppose that we observe the random process $\{\xi_t\}$, $t \geqq 0$, which allows the stochastic differential

$$d\xi_t = r\chi(t-\theta)\,dt + \sigma\, dw_t, \qquad \xi_0(\omega) \equiv 0, \tag{4.145}$$

where

$$\sigma > 0, \quad r \neq 0, \quad \chi(t) = \begin{cases} 1, & t \geqq 0, \\ 0, & t < 0. \end{cases}$$

By analogy with the case of discrete time we investigate the problem of detecting the time $\theta = \theta(\omega)$ in the Bayesian and variational formulations.

2. *Bayesian formulation.* Let

$$\mathscr{F}_t = \sigma\{\omega, \pi: \pi, \xi_s(\omega), s \leqq t\},$$
$$\rho(\pi) = \inf_{\tau \in \mathfrak{M}} \{\mathbf{P}_\pi(\tau < \theta) + c\mathbf{E}_\pi(\tau - \theta \,|\, \tau \geqq \theta)\,\mathbf{P}_\pi(\tau \geqq \theta)\}, \tag{4.146}$$

where the constant $c > 0$. As in § 3, the time $\tau^* = \tau^*(\omega, \pi) \in \mathfrak{M}$ is said to be Bayesian if for all $0 \leqq \pi \leqq 1$

$$\mathbf{P}_\pi(\tau^* < \theta) + c\mathbf{E}_\pi(\tau^* - \theta \,|\, \tau^* \geqq \theta)\,\mathbf{P}_\pi(\tau^* \geqq \theta) = \rho(\pi).$$

THEOREM 10. *A Bayesian time* $\tau^* = \tau^*(\omega, \pi)$ *exists and is given by the formula*

$$\tau^*(\omega, \pi) = \inf\{t \geqq 0: \pi_t(\omega, \pi) \geqq A^*\}, \tag{4.147}$$

where

$$\pi_t(\omega, \pi) = \mathbf{P}_\pi(\theta(\omega) \leqq t \,|\, \mathscr{F}_t). \tag{4.148}$$

The threshhold A^* *is the (unique) root of the equation*

$$C^{-1} = \int_0^{A^*} e^{-\Lambda[H(A^*) - H(x)]} \frac{dx}{x(1-x)^2}, \tag{4.149}$$

where $C = c(r^2/2\sigma^2)^{-1}$, $\Lambda = \lambda(r^2/2\sigma^2)^{-1}$ *and* $H(x) = \ln (x/(1-x)) - (1/x)$.
Moreover,

$$\rho(\pi) = \begin{cases} (1 - A^*) + C \int\limits_{1/A^*}^{1/\pi} \frac{e^{\Lambda x}(x-1)^{\Lambda}}{x_2} \left[\int\limits_x^{\infty} \frac{e^{-\Lambda u} u}{(u-1)^{2+\Lambda}} du \right] dx, & 0 \leqq \pi \leqq A^*, \\ 1 - \pi, & A^* \leqq \pi \leqq 1. \end{cases} \tag{4.150}$$

3. For the proof of this theorem we need

LEMMA 11. *For each* $0 \leqq \pi \leqq 1$ *the random function*

$$\Pi_\pi = \{\pi_t(\omega, \pi), \mathscr{F}_t, \mathbf{P}_\pi\} \qquad t \geqq 0,$$

is Markov with differential

$$d\pi_t = \lambda (1 - \pi_t) dt + (r/\sigma)\pi_t(1 - \pi_t)\, d\bar{w}_t, \qquad \pi_0 = \pi, \tag{4.151}$$

where $\bar{w}_t$ *is a standard Wiener process.*

PROOF. By Lemma 4, for each π, $0 \leqq \pi \leqq 1$, there is a standard Wiener process $\{\bar{w}_t\}$, $t \geqq 0$, such that

$$d\xi_t = r\pi_t\, dt + \sigma d\bar{w}_t, \qquad \xi_0(\omega) \equiv 0. \tag{4.152}$$

We let $\mu_s(\cdot)$, $s \geqq 0$, denote the measure in the space of real functions $x = \{x_t\}$ $t \geqq 0$, $x_0 = 0$, corresponding to the process $\{\eta_t^s\}$, $t \geqq 0$, with stochastic differential

$$d\eta_t^s = r\chi(t - s)\, dt + \sigma\, d\bar{w}_t, \qquad \sigma > 0, \quad \eta_0(\omega) \equiv 0.$$

In the case $s = \infty$ we let $\mu(\cdot) = \mu_\infty(\cdot)$.
Let

$$\nu_t^\pi(\cdot) = \pi\mu_0(\cdot) + (1 - \pi) \int\limits_0^t \lambda e^{-\lambda s} \mu_s(\cdot)\, ds$$

and

$$\nu^\pi(\cdot) = \nu_\infty^\pi(\cdot).$$

From Bayes' formula it follows that

$$\pi_t(\omega, \pi) = \mathbf{P}_\pi\{\theta(\omega) \leqq t \mid \xi_0^t\} = (d\nu_t^\pi/d\nu^\pi)(\xi_0^t),$$

where $(d\nu_t^\pi/d\nu^\pi)(\xi_0^t)$ is the Radon-Nikodým derivative of the measure ν_t^π with respect to the measure ν^π at the "point" ξ_0^t (see **[67]**). Since the measures ν_t^π, ν^π and μ are mutually absolutely continuous, then (ν^π-a. s.)

$$\frac{d\nu_t^\pi}{d\nu^\pi}(\xi_0^t) = \frac{d\nu_t^\pi}{d\mu}(\xi_0^t) \bigg/ \frac{d\nu_t}{d\mu}(\xi_0^t)$$

$$= \frac{\pi \dfrac{d\mu_0}{d\mu}(\xi_0^t) + (1-\pi)\int_0^t \lambda e^{-\lambda s} \dfrac{d\mu_s}{d\mu}(\xi_0^t)\, ds}{\pi \dfrac{d\mu_0}{d\mu}(\xi_0^t) + (1-\pi)\int_0^t \lambda e^{-\lambda s} \dfrac{d\mu_s}{d\mu}(\xi_0^t)\, ds + (1-\pi)\int_t^\infty \lambda e^{-\lambda s} \dfrac{d\mu_s}{d\mu}(\xi_0^t)\, ds}$$

But it is well known [67] that (μ-a.s.)

$$\frac{d\mu_s}{d\mu}(\xi_0^t) = \exp\left\{\frac{r}{\sigma^2}\left[(\xi_t - \xi_s) - \frac{r}{2}(t-s)\right]\right\}, \qquad 0 \leqq s \leqq t,$$

and obviously $(d\mu_s/d\mu)(\xi_0^t) \equiv 1$, $s > t$. Therefore (ν -a.s.)

$$\pi_t(\omega,\pi) = \frac{\pi e^{(r/\sigma^2)[\xi_t - (r/2)t]} + (1-\pi)\int_0^t e^{(r/\sigma^2)[(\xi_t-\xi_s)-(r/2)(t-s)]}\lambda e^{-\lambda s} ds}{\pi e^{(r/\sigma^2)[\xi_t - (r/2)t]} + (1-\pi)\int_0^t e^{(r/\sigma^2)[(\xi_t-\xi_s)-(r/2)(t-s)]}\lambda e^{-\lambda s}\, ds + (1-\pi)e^{-\lambda t}} \tag{4.153}$$

and consequently

$$1 - \pi_t(\omega,\pi) = \frac{(1-\pi)e^{-\lambda t}}{\pi e^{(r/\sigma^2)[\xi t - (r/2)t]} + (1-\pi)\int_0^t e^{(r/\sigma^2)[(\xi_t-\xi_s)-(r/2)(t-s)]}\lambda e^{-\lambda s} ds + (1-\pi)e^{-\lambda t}} \tag{4.154}$$

By Kolmogorov's inequality [22],

$$\mathbf{P}_\pi\left\{\sup_{s\leqq t} |\bar{w}_s| \geqq c\right\} \leqq \frac{t}{c^2}, \qquad t>0.$$

Therefore $\mathbf{P}_\pi\{\sup_{s\leqq t} |\bar{w}_s| < \infty\} = 1$, $t \geqq 0$, and according to (4.152)

$$\mathbf{P}_\pi\left\{\sup_{s\leqq t} |\xi_s| < \infty\right\} = 1. \tag{4.155}$$

From (4.155) it follows that for all finite $t \geqq 0$

$$\mathbf{P}_\pi\{\pi_s < 1; s \leqq t\} = 1, \tag{4.156}$$

provided $0 \leqq \pi < 1$.

Let $\varphi_t = \pi_t/(1-\pi_t)$. By (4.156) we have

$$\mathbf{P}_\pi\{\varphi_s < \infty; s \leqq t\} = 1, \qquad 0 \leqq \pi < 1, \quad t \geqq 0, \tag{4.157}$$

and, according to (4.153) and (4.154), for all $0 \leqq \pi < 1$

$$\varphi_t = \frac{\pi}{1-\pi} e^{\lambda t} . e^{(r/\sigma^2)[\xi_t - (r/2)t]} + \lambda e^{\lambda t} \int_0^t e^{(r/\sigma^2)[(\xi_t - \xi_s) - (r/2)(t-s)]} e^{-\lambda s} \, ds. \tag{4.158}$$

Differentiating (4.158) stochastically, we easily find that for all $0 \leqq \pi < 1$

$$d\varphi_t = \lambda(1+\varphi_t)\,dt + \varphi_t\, d\xi_t, \qquad \varphi_0(\omega, \pi) = \pi/(1-\pi). \tag{4.159}$$

Hence for $\pi_t = \varphi_t/(1+\varphi_t)$ by Itô's formula for change of variable ([**25**], Theorem 7.2; [**30**], p. 387), which is applicable here we obtain from (4.157),

$$d\pi_t = (1-\pi_t)(\lambda - (r^2/\sigma^2)\pi_t^2)\,dt + (r/\sigma^2)\pi_t(1-\pi_t)\,d\xi_t. \tag{4.160}$$

Taking (4.152) into account, from (4.160) we arrive at the representation (4.151).

4. PROOF OF THEOREM 10. As in the case of discrete time with a family of Markov random functions $\{\Pi_\pi, 0 \leqq \pi \leqq 1\}$, in the natural fashion we associate the Markov process $\Pi' = (\pi_t(\omega'), \mathscr{F}'_t, \mathbf{P}'_\pi)$, $t \in \boldsymbol{T}$, where $\omega' = (\omega, \pi)$, $\pi_0(\omega, \pi) = \pi$, $\mathscr{F}'_t = \sigma\{\omega : \pi_0(\omega'), \xi_s(\omega'), s \leqq t\}$, $\xi_s(\omega') = \xi_s(\omega)$ and the measure $\mathbf{P}'_\pi$ can be constructed analogously to (4.15).

Let $\mathfrak{M}^\pi$ be the class of Markov times $\tau \in \mathfrak{M}$ such that $\{\omega' : \tau(\omega') \leqq t\} \in \mathscr{F}^\pi_t$, where $\mathscr{F}^\pi_t = \sigma\{\omega' : \pi_s(\omega'), s \leqq t\}$. As in § 3, it can be established that

$$\begin{aligned}\rho(\pi) &= \inf_{\tau \in \mathfrak{M}^\pi} \mathbf{E}'_\pi \left\{ (1-\pi_\tau) + c\int_0^\tau \pi_s\, ds \right\} \\ &= \inf_{\tau \in \mathfrak{N}^\pi} \mathbf{E}'_\pi \left\{ (1-\pi_\tau) + c\int_0^\tau \pi_s\, ds \right\},\end{aligned} \tag{4.161}$$

where $\mathfrak{N}^\pi$ is the class of Markov times $\tau \in \mathfrak{M}^\tau$ for which

$$\mathbf{E}'_\pi \int_0^\tau \pi_s\, ds < \infty, \qquad 0 \leqq \pi \leqq 1. \tag{4.162}$$

Since

$$\mathbf{E}'_\pi \int_0^\tau \pi_s\, ds = \mathbf{E}_\pi \max(\tau - \theta, 0)$$

and $\mathbf{E}_\pi\, \theta < \infty$, $0 \leqq \pi \leqq 1$, the class $\mathfrak{N}^\pi = \{\tau\}$ coincides with the class of Markov times $\mathfrak{M}^* = \{\tau\}$ for which $\mathbf{E}'_\pi\, \tau < \infty$, $0 \leqq \pi \leqq 1$.

Let

$$g(\pi) = 1 - \pi,$$

$$Q_{c,n} g(\pi) = \min\left\{ g(\pi), \mathbf{E}'_\pi \left[g(\pi_{2^{-n}}) + c\int_0^{2^{-n}} \pi_s\, ds \right] \right\},$$

and suppose $Q^N_{c,n}$ is the Nth power of the operator $Q_{c,n}$.

It is obvious that

$$\mathbf{E}'_\pi \int_0^\Delta \pi_s\, ds = \pi\Delta + (1-\pi)\int_0^\Delta \mathbf{P}_\pi\{\theta \leqq s \mid \theta > 0\} ds$$
$$= \Delta - \frac{1-\pi}{\lambda}(1 - e^{-\lambda\Delta}), \qquad \Delta \geqq 0.$$

Therefore

$$Q_{c,n} g(\pi) = \min\left\{ g(\pi), c\left[\Delta - \frac{1-\pi}{\lambda}(1 - e^{-\lambda\Delta})\right] + \mathbf{E}'_\pi g(\pi_\Delta)\right\},$$

where $\Delta = 2^{-n}$.

The operator T_Δ preserves the convexity of functions. Hence it follows that each of the functions $Q_{c,n} g(\pi)$ and $Q^N_{c,n} g(\pi)$ is convex.

In the problem now being considered the fee is nonzero. Nevertheless, the theory developed in Chapter III for finding optimal stopping rules is applicable. Indeed, just as we did in investigating the optimal stopping problem in § 2 of this chapter, we can show that the given problem can be reduced to a problem in which the fee is equal to zero. [As the function $f(\pi)$ satisfying the relationship

$$\mathbf{E}_\pi f(\pi_\tau) - f(\pi) = \mathbf{E}_\pi \int_0^\tau c\pi_s\, ds$$

for all $\tau = \tau(\omega)$ such that $\mathbf{E}_\pi \tau < \infty$, $0 \leqq \pi \leqq 1$, we can take, for example, the function $f(\pi) = -\int_0^\pi \psi^*(x)\, dx$, where $\psi^*(x)$ is defined below in (4.172).]

According to § 3.8 of Chapter III,

$$\rho(\pi) = \lim_{n\to\infty} \lim_{N\to\infty} Q^N_{c,n} g(\pi).$$

Thus the function $\rho(\pi)$ is always convex and continuous on the interval (0,1).

Since

$$\lim_{t\to\infty} \int_0^t \pi_s\, ds = \infty \qquad (\mathbf{P}_\pi\text{-a.s., } 0 \leqq \pi \leqq 1) \tag{4.163}$$

and $c > 0$, by Remark 1 of Theorem III.5 an optimal time $\tau^* = \tau^*(\omega')$ in the class $\mathfrak{R}^\pi$ exists and is given by the formula

$$\tau^*(\omega') = \inf\{t \geqq 0 : \rho(\pi_t) = g(\pi_t)\} = \inf\{t \geqq 0 : \pi_t \geqq A^*\}. \tag{4.164}$$

We now concern ourselves with finding the unknown constant A^* and the risk $\rho(\pi)$.

To this end, we first assume that in a neighborhood of the point A^* the function $\rho(\pi)$ has a continuous derivative $d\rho(\pi)/d\pi$. Then (see Remark 1 to Theorem III.7, and Theorem III.8)

$$\begin{aligned} &\mathfrak{A}\rho(\pi) = -c\pi, \quad 0 \leqq \pi < A^*, \\ &\rho(\pi) = g(\pi), \quad A^* \leqq \pi \leqq 1, \\ &\left.\frac{d\rho(\pi)}{d\pi}\right|_{\pi=A^*} = \left.\frac{dg(\pi)}{d\pi}\right|_{\pi=A^*}. \end{aligned} \tag{4.165}$$

Suppose $F^* = \{f(\pi)\}$, $0 \leqq \pi \leqq 1$, is the class of nonnegative, convex, twice continuously differentiable functions. We first seek a solution of problem (4.165) in the class F^*. If $f \in F^*$, then $\mathfrak{A}f(\pi) = \mathfrak{D}f(\pi)$ (see Theorem 5.7 in [25]), where by (4.151)

$$\mathfrak{D} = \lambda(1-\pi)\, d/d\pi + (r^2/2\sigma^2)\,[\pi(1-\pi)]^2\, d^2/d\pi^2. \tag{4.166}$$

Consequently problem (4.165) reduces to finding functions $f \in F^*$ and constants A, $0 \leqq A \leqq 1$, for which the following conditions are satisfied:

$$\lambda(1-\pi)f'(\pi) + \rho[\pi(1-\pi)]^2 f''(\pi) = -c\pi, \quad 0 \leqq \pi < A, \tag{4.167}$$

$$f(\pi) = 1-\pi, \quad A \leqq \pi \leqq 1, \tag{4.168}$$

$$f'(A) = -1, \tag{4.169}$$

where $\rho = r^2/2\sigma^2$.

The two conditions (4.168) and (4.169) are still not sufficient for a unique solution of this problem, since the general solution of equation (4.167) contains two undetermined constants and the point A is also unknown.

We shall show that a solution $f^* = f^*(\pi)$ of problem (4.167)–(4.169) exists in the class F^* and is unique, and

$$\left.\frac{df^*(\pi)}{d\pi}\right|_{\pi=0} = 0. \tag{4.170}$$

Let $C = c/\rho$, $\Lambda = \lambda/\rho$ and $\psi(\pi) = f'(\pi)$. From (4.167) we find that

$$\psi'(\pi) = -(C\pi + \Lambda(1-\pi)\psi(\pi))/[\pi(1-\pi)]^2. \tag{4.171}$$

This equation has the singular point $\pi = 0$ and separatrix $\psi^*(\pi)$ at this point ($\psi^*(0) = 0$). It is easy to find that

$$\psi^*(\pi) = -C\int_0^\pi e^{-\Lambda[H(\pi)-H(y)]}\frac{dy}{y(1-y)^2}, \tag{4.172}$$

where $H(y) = \ln(y/(1-y)) - 1/y$.

Let A^* be the root of the equation

$$\psi^*(A^*) = -1 \tag{4.173}$$

and

$$f^*(\pi) = \begin{cases} (1 - A^*) - \int_{\pi}^{A^*} \psi^*(x)\, dx, & 0 \leqq \pi < A^*, \\ 1 - \pi, & A^* \leqq \pi \leqq 1. \end{cases} \tag{4.174}$$

The function $f^* = f^*(\pi)$ is nonnegative and convex, and is a solution of problem (4.167)–(4.169). We shall show that this solution is unique in the class F^*.

To this end we consider the family of integral curves of equation (4.171).

Let the point $A > A^*$ and suppose $\psi_A(\pi)$ is a solution of this equation satisfying condition (4.169), i.e. suppose $\psi_A(A) = -1$. Then $\psi_A(0) = +\infty$, and consequently the solution $f_A(\pi)$ of equation (4.167) such that $f_A(A) = 1 - A$ and $f'_A(A) = \psi_A(A) = -1$ is not a convex function.

Now let the point $B < A^*$ and suppose $\psi_B(\pi)$ is a solution of problem (4.171) satisfying the condition $\psi_B(B) = -1$. Then $\psi_B(0) = -\infty$ and it can easily be seen that a solution $f_B(\pi)$ of equation (4.167) satisfying the conditions $f_B(B) = 1 - B$ and $f'_B(B) = \psi_B(B) = -1$ is such that $f_B(0) < 0$.

Consequently the solution $(A^*, f^*(\pi))$ of problem (4.167)–(4.169) exists, is unique and is determined by (4.173) and (4.174).

We shall now show that the function $f^*(\pi)$ that we found coincides with the risk $\rho(\pi)$. For this, we use the same approach that we used for the proof of the analogous assertion in Theorem 2.

If $\pi < A^*$, then according to the Corollary to Theorem 5.1 in [25], for every $\tau \in \mathfrak{R}^\pi$

$$\mathbf{E}'_\pi f^*(\pi_\tau) - f^*(\pi) = -c\mathbf{E}_\pi \int_0^\tau \pi_s\, ds.$$

Then for $\pi < A^*$(cf. (4.62))

$$\begin{aligned} \rho(\pi) &= \inf_{\tau \in \mathfrak{R}^\pi} \mathbf{E}'_\pi \left\{ g(\pi_\tau) + c\int_0^\tau \pi_s\, ds \right\} \\ &\geqq \inf_{\tau \in \mathfrak{R}^\pi} \mathbf{E}'_\pi \left\{ f^*(\pi_\tau) + c\int_0^\tau \pi_s\, ds \right\} + \inf_{\tau \in \mathfrak{R}^\pi} \mathbf{E}'_\pi \{ g(\pi_\tau) - f^*(\pi_\tau) \} \\ &= f^*(\pi) + \inf_{\tau \in \mathfrak{R}^\pi} \mathbf{E}'_\pi \{ g(\pi_\tau) - f^*(\pi_\tau) \}. \end{aligned} \tag{4.175}$$

But $g(\pi) \geqq f^*(\pi)$ for all $0 \leqq \pi \leqq 1$, and for the time $\tau^* = \inf\{t \geqq 0\colon \pi_t \geqq A^*\}$, obviously

$$\mathbf{E}'_\pi \{ g(\pi_{\tau^*}) - f^*(\pi_{\tau^*}) \} = 0.$$

Consequently $\rho(\pi) \geqq f^*(\pi)$. But

$$\mathbf{E}'_\pi\left\{ g(\pi_{\tau^*}) + c\int_0^{\tau^*} \pi_s ds \right\}$$

$$= \mathbf{E}'_\pi\left\{ f(\pi_{\tau^*}) + c\int_0^{\tau^*} \pi_s ds \right\} = f^*(\pi), \qquad 0 \leqq \pi < A^*.$$

Thus for all $0 \leqq \pi < A^*$ the risk $\rho(\pi)$ coincides with $f^*(\pi)$. Analogously we can also verify the equality $\rho(\pi) = f^*(\pi)$ for $\pi \geqq A^*$, which completes the proof of Theorem 10.

5. Knowledge of the Bayesian time $\tau^*(\omega, \pi)$ makes it easy here to find the optimal time $\bar{\tau}$ in the variational formulation as well.

THEOREM 11. *Let* $0 < \alpha < 1, 0 \leqq \pi < 1$ *and* $0 < \lambda < \infty$, *and suppose* $\mathfrak{M}(\alpha;\pi)$ *is the collection of Markov times for which* $\mathbf{P}_\pi(\tau < \theta) \leqq \alpha$. *Then the Markov time*

$$\bar{\tau}_\alpha = \inf\{t \geqq 0: \pi_t(\omega, \pi) \geqq \tilde{A}_\alpha\}, \tag{4.176}$$

where

$$\tilde{A}_\alpha = 1 - \alpha, \tag{4.177}$$

is optimal in the sense that for every $\tau \in \mathfrak{M}(\alpha;\pi)$

$$\mathbf{E}_\pi(\bar{\tau}_\alpha - \theta \,|\, \bar{\tau}_\alpha \geqq \theta) \leqq \mathbf{E}_\pi(\tau - \theta \,|\, \tau \geqq \theta).$$

The proof of this theorem can be carried out in the same way as the proof of Theorem 9. We only note that the equality $\tilde{A}_\alpha = 1 - \alpha$ follows from the fact that for all $\pi \leqq \tilde{A}_\alpha$

$$\mathbf{E}_\pi(1 - \pi_{\bar{\tau}_\alpha}) = 1 - \tilde{A}_\alpha,$$

and if $\pi > \tilde{A}_\alpha$, then

$$\mathbf{E}_\pi(1 - \pi_{\bar{\tau}_\alpha}) = 1 - \pi.$$

6. We now turn to the question of what is the mean delay time

$$R(\alpha;\lambda) = \mathbf{E}_0(\bar{\tau}_\alpha - \theta \,|\, \bar{\tau}_\alpha \geqq \theta) \tag{4.178}$$

for given probability of false alarm α (restricting ourselves to the case $\pi_0 = 0$ for simplicity).

Let c_α be that constant c involved in (4.146) for which the Bayesian time $\tau^*(\omega,\pi)$ (see (4.147)) coincides with the time $\bar{\tau}_\alpha$ defined in (4.176). (The existence of c_α follows from reasoning analogous to that used in the proof of Theorem 9.)

Then, according to (4.150),

$$\rho(0)= \alpha + \frac{c_\alpha}{\rho}\int_{1/A_\alpha}^{\infty}\frac{e^{\Lambda x}(x-1)^{\Lambda}}{x^2}\left[\int_x^{\infty}\frac{e^{-\Lambda u}u\,du}{(u-1)^{2+\Lambda}}\right]dx. \tag{4.179}$$

On the other hand,

$$\begin{aligned}\rho(0) &= \inf_{\tau\in\mathfrak{M}}\{\mathbf{P}_0(\tau<\theta)+c_\alpha\mathbf{P}_0(\tau\geqq\theta)\,\mathbf{E}_0(\tau-\theta\mid\tau\geqq\theta)\}\\ &=\mathbf{P}_0(\bar{\tau}_\alpha<\theta)+c_\alpha\,\mathbf{P}_0(\tau_\alpha\geqq\theta)\,R(\alpha;\lambda)=\alpha+c_\alpha(1-\alpha)\,R(\alpha;\lambda).\end{aligned} \tag{4.180}$$

Comparing (4.179) and (4.180), we find

$$R(\alpha;\lambda)=\frac{\displaystyle\int_{1/(1-\alpha)}^{\infty}\frac{e^{\Lambda x}(x-1)^{\Lambda}}{x^2}\left[\int_x^{\infty}\frac{e^{-\Lambda u}u}{(u-1)^{2+\Lambda}}\,du\right]}{(1-\alpha)\rho}. \tag{4.181}$$

We examine this formula in the case of greatest practical interest, $\lambda\to 0$. Naturally, as $\lambda\to 0$, i.e. when the mean time of appearance of a disruption $\mathbf{E}_0\,\theta=1/\lambda$ approaches infinity, it is reasonable to assume that also $\alpha\to 1$. We assume that $\lambda\to 0$ and $\alpha\to 1$, but in such a way that the ratio $(1-\alpha)/\lambda=T$, where T is fixed.

Then from (4.181), as $\alpha\to 1$ and $\lambda\to 0$, and for a fixed ratio $(1-\alpha)/\lambda=T$, we find

$$\begin{aligned}R(T)&=\lim R(\alpha;\lambda)=\lim\frac{\displaystyle\Lambda\int_{\Lambda/(1-\alpha)}^{\infty}\frac{e^y}{y^2}\left(\int_y^{\infty}\frac{e^{-z}}{z}dz\right)dy}{\rho(1-\alpha)}\\ &=\frac{1}{\rho}\lim\frac{1}{\Lambda-\alpha}\int_{\Lambda/(1-\alpha)}^{\infty}\frac{e^y}{y^2}\left(\int_y^{\infty}\frac{e^{-z}}{z}dx\right)dy\\ &=\frac{b}{\rho}\int_b^{\infty}\frac{e^y}{y^2}[-\operatorname{Ei}(-y)]\,dy,\end{aligned} \tag{4.182}$$

where $b=(\rho T)^{-1}$ and $-\operatorname{Ei}(-y)=\int_y^{\infty}(e^{-z}/z)dz$ is the exponential integral function.

After simple transformations [60] we obtain

$$\begin{aligned}R(T)&=\frac{b}{\rho}\int_b^{\infty}e^y\operatorname{Ei}(-y)\,d\left(\frac{1}{y}\right)\\ &=\frac{b}{\rho}\left\{-\frac{1}{b}e^b\operatorname{Ei}(-b)-\int_b^{\infty}\frac{e^z}{z}\operatorname{Ei}(-z)\,dz-\int_b^{\infty}\frac{dz}{z^2}\right\}\\ &=\frac{1}{\rho}\left\{e^b[-\operatorname{Ei}(-b)]-1+b\int_b^{\infty}\frac{e^z}{z}[-\operatorname{Ei}(-z)]dz\right\}.\end{aligned}$$

But

$$-\operatorname{Ei}(-z)=e^{-z}\int_0^\infty \frac{e^{-y}}{y+z}\,dy,$$

whence

$$\int_b^\infty \frac{e^z}{z}[-\operatorname{Ei}(-z)]\,dz=\int_0^\infty e^{-y}\left(\int_b^\infty \frac{dz}{z(z+y)}\right)dy$$

$$=\int_0^\infty \frac{e^{-z}\ln(1+z/b)}{z}\,dz.$$

Consequently

$$R(T)=\frac{1}{\rho}\left[e^b[-\operatorname{Ei}(-b)]-1+b\int_0^\infty e^{-bz}\ln((1+z)/z)\,dz,\right], \tag{4.183}$$

where $b=(\rho T)^{-1}$.

In the case of large T, from (4.183) we obtain [**60**]

$$R(T)=(1/\rho)\{\ln(\rho T)-1-\boldsymbol{C}+O(1/\rho T)\},$$

where $\boldsymbol{C}=0.577\ldots$ is Euler's constant.

Notes

Chapter I

§ 1. The axiomatics of probability theory are presented in the fundamental work by Kolmogorov [**41**]. Proofs of the results given concerning measurability of random processes are contained in the monographs by Dynkin [**25**] and Meyer [**49**].

§ 2. Additional information about the properties of Markov times can be found in the monographs of Dynkin [**25**], Meyer [**49**] and [**50**], and Blumental and Getoor [**12**]. Theorem 1 is due to Courrège and Priouret [**20**]. The Remark to Theorem 1 was made by Hans-Jürgen Engelbert.

§ 3. The basic definitions given for the theory of Markov processes follow the monographs of Dynkin [**24**] and [**25**], and Blumental and Getoor [**12**].

§ 4. Proofs of the given theorems on martingales and super-martingales are given in Doob [**22**] and Meyer [**49**]. Generalized martingales and super-martingales are investigated in an article by Snell [**68**].

Chapter II

§ 1. The value $s(x)$ for the case of nonnegative functions was considered in the article by Dynkin [**26**], where he also gave properties of excessive functions and Lemma 2. A somewhat different proof of Lemma 1 is given in Meyer [**49**]. Lemma 3 is contained in work by Grigelionis and Širjaev [**37**]. The method for constructing a smallest excessive majorant, given in Lemma 4, was presented by A.D. Ventcel′ (see [**27**], Chapter III, Problem 21). In martingale theory Lemma 5 is known under the name of the theorem on transformation under a system of optional sampling (Doob [**22**], Theorem 2.2 in Chapter VII). The proof of Theorem 6 was adapted from Snell [**68**]. The method for constructing the s.e.m. of the function $g(x)$, given in Lemmas 7 and 9, is presented for the first time. Similar constructions are also contained in the article by Siegmund [**56**]. Another method of proof of Theorem 1, for the case $g(x) \geqq 0$, was given in the article by Dynkin [**26**].

§ 2. $(\varepsilon,\bar{s})$-optimal times are investigated for the first time. In the case $0 \leqq g(x) \leqq C < \infty$, (ε,s)-optimality of the time τ_ε was proved by Dynkin [**26**]. Assertions 2) and 3) of Theorem 2 are close to results in articles by Siegmund [**56**] and Chow and Robbins [**17**]. The problem of choosing the best object, also known as the "fastidious bride" problem, was investigated by Gardner [**29**] and Dynkin [**26**] (see also [**27**]). Similarly formulated problems were considered in articles by Chow,

Moriguti, Robbins and Samuels [19], Gilbert and Mosteller [32], and Guseĭn-Zade [38].

§ 3. The example presented at the beginning of this section was given by Haggstrom [39]. The classes of Markov times $\mathfrak{N}$ and $\mathfrak{R}$ were considered in the article by Chow and Robbins [17]; their methods were also used in proving Lemma 10 and Theorems 3 and 4.

§ 4. Theorems 5 and 6 are contained in the work of Chow and Robbins [15], [16], [17], and Haggstrom [39].

§ 5. Questions of uniqueness of the solution of the recursion equations (2.85) were investigated by Bellman [9], Grigelionis and Širjaev [37] and Grigelionis [35]. Theorem 10 is due to Siegmund [56].

§ 6. The results given in this section were obtained in the article by Ray [53], and also by Grigelionis and Širjaev.

§ 7. Randomized and sufficient classes of Markov times were investigated in works by Siegmund [56], Širjaev [62], Dynkin [28], and Grigelionis [36].

§ 8. Functionals of the type (2.113) were studied in the article by Krylov [43]. The example given at the end of the section for the case $\alpha = 1$ was considered in articles by Chow and Robbins [15], [16] and Siegmund [56].

Chapter III

§ 1. The definitions and proofs of the given properties of excessive functions are due to Hunt [40] and Dynkin [25] (see also [12] and [49]).

§ 2. The method for constructing the smallest excessive majorant of the function $g(x)$ (Lemma 1) was given by Grigelionis and Širjaev [37]. Another method was presented earlier by Dynkin [26].

§ 3. Theorem 1 for the case $g(x) \geqq 0$ was presented by Dynkin [26]. The value $\bar{s}(x)$ for continuous Markov processes has not been investigated previously. The example given in § 3.3 is contained in the article by Taylor [57]. In the proof of Lemma 8 we used methods from the article by Dynkin [26]. Lemma 9 and Theorems 2 and 3 were obtained by the author.

§ 4. Assertion 1 of Theorem 4 was proved by Dynkin [26]. Results close to assertions 2) and 3) were also obtained by Siegmund [56]. Theorem 5 is given for the first time. Under special assumptions, Theorem 6 was proved by Taylor [70].

§ 5. Lemmas 11 and 12 are published for the first time. Theorems 7, 8 and 9 were obtained by Grigelionis and Širjaev [37].

§ 6. The "smooth pasting" conditions was used to solve concrete problems in work by Mihalevič [51], Chernoff [14], Lindley [46], Bather [6], Širjaev [61], Whittle [76] and Stratonovič [69]. Theorem 10 is due to Grigelionis and Širjaev [37]. Theorems 12 and 13 were obtained by Grigelionis [34].

Chapter IV

§ 1. The Bayesian and variational formulations of the problems of sequential testing of two simple hypotheses are due to Wald [**72**]. The properties of transitive statistics were investigated in work by Bahadur [**4**], Širjaev [**62**], [**65**], and Grigelionis [**36**].

Proofs of Lemma 3 and Theorem 1 are contained in the articles by Chow and Robbins [**16**] and Širjaev [**66**].

Equations (4.36)–(4.37) were first obtained by Mihalevič [**51**]. Lemma 4 is contained in the article by Širjaev [**64**] (see also [**47**]). A somewhat different proof of Theorem 2 is given by Širjaev [**66**]. The idea of the proof of the equality $\rho(\pi) = f^*(\pi)$ used by us is due to B. Rozovskiĭ.

§ 2. Theorem 3 is due to Wald [**72**]. Lemma 8 and its proof were given by Shepp [**55**]. He is also credited with the results presented in Remarks 3 and 4 to this lemma. The comparison of the optimal properties of the Neyman-Pearson method and the sequential probability ratio test was given by Aĭvazjan [**1**]. An elegant proof of Theorem 4 is contained in the book by Lehmann [**45**]. The bounds (4.98) contained in Theorem 5 were presented by Wald [**72**]. Theorem 6 was obtained by Stein (see [**72**]).

Wald's identity (Lemma 10) has been the subject of investigation by many authors: Wald [**72**], Blackwell [**10**], Doob [**22**], Chow, Robbins and Teicher [**18**], Shepp [**55**], and others.

Theorem 7 in the case $N = 2$ was obtained by Wald [**72**]. In the general case it was obtained by Hoeffding (communicated by him to the author in 1965). The proof given here is contained in the article by Simons [**57**]. An analogous proof is presented in the book by Bechhofer, Kiefer and Sobel [**8**] (Theorem 3.5.1), where formula (4.117) is also given.

§ 3. The disruption problem was first investigated in a report by A.N. Kolmogorov and the author at the Sixth All-Union Conference on Probability Theory and Mathematical Statistics (Vilnius, 1960). The results are contained in [**58**], [**60**] and [**62**].

§ 4. The disruption problem for a Wiener process was investigated by Širjaev in [**59**],[**60**], [**63**] and [**66**]. Other formulations of problems of quickest detection of the time of appearance of a disruption were also considered in these works. The disruption problem was also analyzed by Stratonovič [**69**] and Bather [**7**].

Bibliography

[1] S. A. Aĭvazjan, *A comparison of the optimal properties of the Neyman-Pearson and the Wald sequential probability ratio tests,* Teor. Verojatnost. i Primenen. **4**(1959), 86–93 = Theor. Probability Appl. **4**(1959), 83–88. MR **21** #942.

[2] P. S. Aleksandrov, *Introduction to the general theory of sets and functions,* GITTL, Moscow, 1948; German transl., VEB Deutscher Verlag, Berlin, 1956. MR **12**, 682; **18,** 22.

[3] K. J. Arrow, D. Blackwell and M. A. Girshick, *Bayes and minimax solutions of sequential decision problems,* Econometrica **17** (1949), 213–214. MR **11,** 261.

[4] R. R. Bahadur, *Sufficiency and statistical decision functions,* Ann. Math. Statist. **25** (1954), 423–462. MR **16,** 154.

[5] A. E. Bašarinov and B. S. Fleĭšman, *Methods of statistical sequential analysis and its application to radio engineering,* Izdat. "Soviet Radio", Moscow, 1962. (Russian)

[6] J. A. Bather, *Bayes procedures for deciding the sign of a normal mean,* Proc. Cambridge Philos. Soc. **58** (1962), 599–620. MR **27** #869.

[7] _____, *On a quickest detection problem,* Ann. Math. Statist. **38** (1967), 711–724. MR **35** #2679.

[8] R. E. Bechhofer, J. Kiefer and M. Sobel, *Sequential identification and ranking procedures, with special reference to Koopman-Darmois populations,* Statistical Research Monographs, vol. 3, Univ. of Chicago Press, Chicago, Ill., 1968, MR **39** #6445.

[9] R. E. Bellman, *Dynamic programming,* Princeton Univ. Press, Princeton, N. J., 1957; Russian transl., IL, Moscow, 1960. MR **19,** 820; **22** #4898.

[10] D. H. Blackwell, *On an equation of Wald,* Ann. Math. Statist. **17** (1946), 84–87. MR **8,** 478.

[11] D. H. Blackwell and M. A. Girshick, *Theory of games and statistical decisions,* Wiley, New York; Chapman & Hall, London, 1954; Russian transl., IL, Moscow, 1958. MR **16,** 1135.

[12] R. M. Blumental and R. K. Getoor, *Markov processes and potential theory,* Pure and Appl. Math. vol. 29, Academic Press, New York and London, 1968, MR **41** #9348.

[13] D. L. Burkholder and R. A. Wijsman, *Optimum properties and admissibility of sequential tests,* Ann. Math. Statist. **34** (1963), 1-17. MR **26** #5668.

[14] H. Chernoff, *Sequential tests for the mean of a normal distribution,* Proc. Fourth Berkeley Sympos. Math. Statist. and Probability, vol. 1, Univ. of California Press, Berkeley, Calif., 1961, pp. 79–91. MR **24** #A1788.

[15] Y. S. Chow and H. Robbins, *A martingale system theorem and applications,* Proc. Fourth Berkeley Sympos. Math. Statist. and Probability, vol. 1, Univ. of California Press, Berkeley, Calif., 1961, pp. 93–104. MR **24** #A2433.

[16] _____, *On optimal stopping rules,* Z. Wahrscheinlichkeitstheorie und Verw. Gebiete **2** (1963), 33–49. MR **28** #698.

[17] _____, *On values associated with a stochastic sequence,* Proc. Fifth Berkeley Sympos. Math. Statist. and Probability, vol. 1: Statistics, Univ. of California Press, Berkeley, Calif., 1967, pp. 427–440. MR **35** #6193.

[18] Y. S. Chow, H. Robbins and H. Teicher *Moments of randomly stopped sums,* Ann. Math. Statist. **36** (1965), 789–799. MR **31** #2777.

[19] Y. S. Chow, S. Moriguti, H. Robbins and S. M. Samuels, *Optimal selection based on relative rank* (*the "secretary problem"*) Israel J. Math. **2** (1964), 81–90. MR **31** #855.

[20] P. Courrège and P. Priouret, *Temps d'arrêt d'une fonction aléatoire*: *Relations d'équivalence*

associeés et propriétés de décomposition, Publ. Inst. Statist. Univ. Paris **14** (1965), 245–274. MR **36** #4640.

[21] J. Dieudonné, *Fondements de l'analyse moderne,* Pure and Appl. Math., vol. 10, Academic Press, New York and London, 1960; Russian transl., "Mir", Moscow, 1964. MR **22** #11074.

[22] J. Doob, *Stochastic processes,* Wiley, New York; Chapman & Hall, London, 1953; Russian transl., IL, Moscow, 1956. MR **15,** 445; **19,** 71.

[23] A. Dvoretzky, J. Kiefer and J. Wolfowitz, *Sequential decision processes with continuous time parameter. Testing hypotheses,* Ann. Math. Statist. **24** (1953), 254–264. MR **14,** 997; 1279.

[24] E. B. Dynkin, *Foundations of the theory of Markov processes,* Fizmatgiz, Moscow, 1959; English transl., Prentice-Hall, Englewood Cliffs, N.J.; Pergamon Press, Oxford, 1961. MR **24** #A1745; #A1746; #A1747.

[25] _____, *Markov processes,* Fizmatgiz, Moscow, 1963; English transl., Die Grundlehren der math. Wissenschaften, Band 121, Academic Press, New York; Springer-Verlag, Berlin, 1965, MR **33** #1886; #1887.

[26] _____, *The optimum choice of the instant for stopping a Markov process,* Dokl. Akad. Nauk SSSR **150** (1963), 238–240 = Soviet Math. Dokl. **4** (1963), 627–629. MR **27** #4278.

[27] _____, *Sufficient statistics for the optimal stopping problem,* Teor. Verojatnost. i. Primenen. **13** (1968), 150–152 = Theor. Probability Appl. **13** (1968), 152–153. MR **36** #7227.

[28] E. B. Dynkin and A. A. Juškevič, *Theorems and problems in Markov processes,* "Nauka", Moscow, 1967; English transl., Plenum Press, New York, 1969. MR **36** #6006; 39 #3585 a.

[29] M. Gardner, *Mathematical games,* Sci. Amer. **202** (1960), no. 2, 150–156; ibid. **202** (1960), no. 3, 173–182.

[30] *I. I. Gihman and A. V. Skorohod, *Introduction to the theory of random processes,* "Nauka", Moscow, 1965; English transl., Saunders, Philadelphia, Pa., 1969. MR **33** #6689; **40** #923.

[31] _____, *Stochastic differential equations,* "Naukova Dumka", Kiev, 1968. (Russian) MR **41** 7777.

[32] J. P. Gilbert and F. Mosteller, *Recognizing the maximum of a sequence,* J. Amer. Statist. Assoc. **61** (1966), 35–73. MR **33** #6792.

[33] L. M. Graves, *The theory of functions of real variables,* McGraw-Hill, New York, 1946. MR **8,** 319.

[34] B. I. Grigelionis, *The optimal stopping of Markov processes,* Litovsk. Mat. Sb. **7** (1967), 265–279; English transl., Selected Transl., Math. Statist. and Probability, vol. 11, Amer. Math. Soc., Providence, R. I. (to appear). MR **38** #2838.

[35] _____, *Conditions for the uniqueness of the solution of Bellman's equation,* Litovsk. Mat. Sb. **8** (1968), 47–52. (Russian) MR **39** #5206.

[36] _____, *Sufficiency in optimal stopping problems,* Litovsk. Mat. Sb. **9** (1969), 471–480. (Russian)

[37] B. I. Grigelionis and A. N. Širjaev, *On the Stefan problem and optimal stopping rules for Markov processes,* Teor. Verojatnost. i Primenen. **11** (1966), 612–631 = Theor. Probability Appl. **11** (1966), 541–558. MR **35** #7538.

[38] S. M. Guseĭn-Zade, *The problem of choice and the optimal stopping rule for a sequence of independent trials,* Teor. Verojatnost. i Primenen. **11** (1966), 534–537 = Theor. Probability Appl. **11** (1966), 472–475. MR **34** #2129.

[39] G. W. Haggstrom, *Optimal stopping and experimental design,* Ann. Math. Statist. **37** (1966), 7–29. MR **33** #3424.

[40] J. A. Hunt, *Markov processes and potentials.* I, II, III, Illinois J. Math. **1** (1957), 44–93, 316–369; ibid. **2** (1958), 153–213; Russian transl., IL, Moscow, 1962. MR **19,** 951; **21** #5824.

[41] A. N. Kolmogorov, *Grundbegriffe der Wahrscheinlichkeitsrechnung,* Springer, Berlin, 1933; Russian transl., ONTI, Moscow, 1936; English transl., Chelsea, New York, 1948. MR **11,** 374; **18,** 155.

[42] A.N. Kolmogorov and S. V. Fomin, *Elements of the theory of functions and of functional analysis,* "Nauka", Moscow, 1968; English transl. of 1st ed., Graylock Press, Albany, N. Y.; Academic Press, New York, 1961. MR **38** #2559.

Editor's note.* All page references to **[30] are to the English translation.

[43] N. V. Krylov, *On optimal stopping of a control circuit,* Optimal Control and Information Theory, Abstracts of Reports at the Seventh All-Union Conf. Theory of Probability and Math Statist. (Tbilisi, 1963), Izdat. Inst. Mat. Akad. Nauk. Ukrain. SSR, Kiev, 1963, pp. 11–15. (Russian)

[44] S. Kullback, *Information theory and statistics,* Wiley, New York, 1959; Russian transl., "Nauka", Moscow, 1967, MR 21 #2325.

[45] E. Lehmann, *Testing statistical hypothesis,* Wiley, New York; Chapman & Hall, London, 1959; Russian transl., "Nauka", Moscow, 1964. MR **21** #6654.

[46] D. V. Lindley, *Dynamic programming and decision theory,* Appl. Statist. **10** (1961), 39–51. MR **23** #A740.

[47] R. Š. Lipcer and A. N. Širjaev *Nonlinear filtering of Markov diffusion processes,* Trudy Mat. Inst. Steklov. **104** (1968), 135–180 = Proc. Steklov Inst. Math. **104** (1968), 163–218. MR **41** #6342.

[48] M. Loève, *Probability theory. Foundations. Random sequences,* University Series in Higher Math., Van Nostrand, Princeton, N. J., 1960; Russian transl., IL, Moscow, 1962. MR **23** A670; **25** #3551.

[49] P. A. Meyer, *Probabilités et potentiel,* Publ. Inst. Math. Univ. Strasbourg, no. 14, Actualités Sci. Indust., no. 1318, Hermann, Paris, 1966. MR **34** #5118

[50] _____, *Processus de Markov: La frontière de Martin,* Lecture Notes in Math., no. 77, Springer-Verlag, Berlin and New York, 1968. MR **39** #7669.

[51] V. S. Mihalevič, *Bayesian choice between two hypotheses for the mean value of a normal process,* Vīsnik Kiïv. Unīv. **1** (1958), no. 1, 101–104. (Ukrainian)

[52] I. P. Natanson, *Theory of functions of a real variable,* 2nd rev. ed., GITTL, Moscow, 1957; English transl., Ungar, New York, 1961. MR **26** #6309

[53] S. N. Ray, *Bounds on the maximum sample size of a Bayes sequential procedure,* Ann. Math. Statist. **36** (1965), 859–878. MR **30** #5446.

[54] L. I. Rubinšteĭn, *The Stefan problem,* Latvian State University Computing Center, Izdat. "Zvaĭgzne", Riga, 1967; English transl., Transl. Math. Monographs, vol. 27, Amer. Math. Soc., Providence, R. I., 1971. MR **36** #5488.

[55] L. A. Shepp, A *first passage problem for the Wiener process,* Ann. Math. Statist. **38** (1967), 1912–1914. MR **36** #968.

[56] D. O. Siegmund, *Some problems in the theory of optimal stopping rules,* Ann. Math. Statist. **38** (1967), 1627–1640. MR **36** #4718.

[57] G. Simons, *Lower bounds for average sample number of sequential multihypothesis tests,* Ann. Math. Statist. **38** (1967), 1343–1364. MR **36** #1037.

[58] A. N. Širjaev, *The detection of spontaneous effects,* Dokl. Akad. Nauk SSSR **138** (1961), 799–801 = Soviet Math. Dokl. **2** (1961), 740–743. MR **24** #A2504.

[59] _____, *The problem of the most rapid detection of a disturbance in a stationary process,* Dokl. Akad. Nauk SSSR **138** (1961), 1039–1042 = Soviet Math. Dokl. **2** (1961), 795–799. MR **24** #A2505.

[60] _____, *Optimal methods in quickest detection problems,* Teor. Verojatnost. i Primenen. **8** (1963), 26–51 = Theor. Probability Appl. **8** (1963), 22–45, MR **27** #5642.

[61] _____, *On the theory of decision functions and control of a process of observation based on incomplete information,* Trans. Third Prague Conf. Information Theory, Statistical Decision Functions, Random Processes (Liblice, 1962), Publ. House Czech. Acad. Sci., Prague, 1964, pp. 657–681; English transl., Selected Transl. Math. Statist. and Probability, vol. 6, Amer. Math. Soc., Providence, R. I., 1966, pp. 162–188. MR **30** #1591.

[62] _____, *On Markov sufficient statistics in non-additive Bayes problems of sequential analysis,* Thor. Verojatnost. i Primenen. **9** (1964), 670–686 = Theor. Probability Appl. **9** (1964), 604–618. MR **30** #350.

[63] _____, *Some explicit formulae in a problem on "disorder",* Teor. Verojatnost. i Primenen. **10** (1965), 380–385 = Theor. Probability Appl. **10** (1965), 348–353. MR **32** #2263.

[64] _____, *Stochastic equations of non-linear filtering of jump-like Markov processes,* Problemy Peredači Informacii **2** (1966), vyp. 3, 3–22. (Russian) MR **34** #4045.

[65] _____, *Some new results in the theory of controlled random processes,* Trans. Fourth Prague

Conf. Information Theory, Statistical Decision Functions, Random Processes (Prague, 1965), Academia, Prague, 1967, pp. 131–203; English transl., Selected Transl. Math. Statist. and Probability, vol. 8, Amer. Math. Soc., Providence, R.I., 1970, pp. 49–130. MR **35** #4066; 37 #1186.

[66] _____, *On two problems of sequential analysis,* Kibernetika (Kiev) **1967**, no. 2, 79–80. (Russian)

[67] A. V. Skorohod, *Studies in the theory of random processes,* Izdat. Kiev. Univ., Kiev, 1961; English transl., Addison-Wesley, Reading, Mass., 1965. MR **32** #3082a,b.

[68] J. L. Snell, *Applications of martingale system theorems,* Trans. Amer. Math. Soc. **73** (1952), 293–312. MR **14**, 295.

[69] R. L. Stratonovič, *Conditional Markov processes and their application to the theory of automatic control,* Izdat. Moskov. Univ., Moscow, 1966; English transl., American Elsevier, New York, 1967. MR **33** #5391.

[70] H. M. Taylor, *Optimal stopping in a Markov process,* Ann. Math. Statist. **39** (1968), 1333–1344. MR **38** #769.

[71] A. Wald, *Statistical decision functions,* Wiley, New York; Chapman & Hall, London, 1950; Russian transl. in *Theory of games*: *Positional games,* "Nauka", Moscow, 1967, pp. 300–522. MR **12**, 193: **36** #4704.

[72] _____, *Sequential analysis,* Wiley, New York; Chapman & Hall, London, 1947; Russian transl., Fizmatgiz. Moscow, 1960. MR **8**, 593; **22** #7228.

[73] A. Wald and J. Wolfowitz, *Optimum character of the sequential probability ratio test,* Ann. Math. Statist. **19** (1948), 326–339 MR **10**, 201.

[74] _____, *Bayes solutions of sequential decision problems,* Ann. Math. Statist. **21** (1950), 82–99. MR **11**, 529.

[75] G. B. Wetherill, *Sequential methods in statistics,* Methuen, London, 1966. MR **35** #7504.

[76] P. Whittle, *Some general results in sequential design* (*with discussion*), J. Roy. Statist. Soc. Ser. B **27** (1965), 371–394. MR **34** #2133.